Lecture Notes in Mobility

Series editor

Gereon Meyer, Berlin, Germany

More information about this series at http://www.springer.com/series/11573

David Beeton · Gereon Meyer
Editors

Electric Vehicle Business Models

Global Perspectives

Springer

Editors
David Beeton
Urban Foresight Ltd.
Newcastle upon Tyne
UK
david.beeton@urbanforesight.org

Gereon Meyer
VDI/VDE Innovation + Technik GmbH
Berlin
Germany
gereon.meyer@vdivde-it.de

ISSN 2196-5544 ISSN 2196-5552 (electronic)
Lecture Notes in Mobility
ISBN 978-3-319-12243-4 ISBN 978-3-319-12244-1 (eBook)
DOI 10.1007/978-3-319-12244-1

Library of Congress Control Number: 2014956874

Springer Cham Heidelberg New York Dordrecht London

Printed on acid-free paper

Springer International Publishing AG Switzerland is part of Springer Science+Business Media (www.springer.com)

Preface

Electric vehicles are a global phenomenon. Global in that every corner of the planet is affected by the environmental, energy security and health imperatives for increased adoption of electric vehicles. Global in that the challenges and opportunities this presents are universal to governments and industry worldwide. Global in that achieving this transformation requires a holistic system of innovative technologies, policies and business models.

This book therefore sets out to explore these global perspectives. It presents insights from around the world and across the network of organisations, technologies, consumers, products and services which characterise the electric vehicle ecosystem. It draws on the networks of the International Energy Agency's Hybrid and Electric Vehicle Implementing Agreement to present commentary and case studies from experts in eleven different countries from across five continents.

The focus on electric vehicle business models recognises that market development will be a key enabler in realising a rapid transition from niche to mainstream adoption. This will demand that electric mobility products and services are developed to provide benefits which exceed apparent costs and supersede any perceived relative advantage of fossil fuelled vehicles. In simple terms, the goal is to make electric vehicles more cost-effective, convenient, desirable and rewarding to use.

History suggests that this is seldom a straightforward process. The right business model is rarely apparent early on in emerging industries. Furthermore, customers and incumbent industry players often face multiple restraining forces and switching costs in adjusting to new technologies or ways of doing business. Therefore, while business model innovation can undoubtedly facilitate greater market adoption of electric vehicles, it also represents an area that is replete with challenges and exposed to continuous change.

This book considers this further and presents a series of discussion papers on electric vehicle business models. This provides expert commentary and analysis from cities, boardrooms and research labs around the world.

It commences by considering the macro dynamics and changes that are at play in the industry as a whole. Electric vehicle business models are placed in a wider context, and conceptual structure is provided for the various solutions and approaches that are emerging in this space.

The next section explores business models for recharging infrastructure. This includes market models and billing strategies for public charge points, the business case for deploying rapid chargers on a motorway network, and solutions for residents of multi-unit dwellings. Consideration is also given to the potential for wireless charging technologies and the associated business models that are emerging for this new technology.

Energy systems are then considered from two perspectives. The first is the potential for electric vehicle batteries to be integrated with grids to provide dynamic storage and supply. The second focuses on the vehicle itself, considering how vehicle design and battery systems influence energy efficiency and the associated total cost of ownership.

Attention then turns to fleet applications of electric vehicles. This reviews the evolution of electric mobility in carsharing business models and the challenges and opportunities that electric vehicles present to carsharing operators around the world. It then presents an example of how fleet managers can use analytical tools to identify applications for electric vehicles to save money and significantly reduce emissions.

The final section presents a series of case studies on different aspects of electric vehicle business models from around the world. This provides lessons learned and conceptual insights from experiences in Japan, China, Hawaii and Chile.

Most new business models emerge from analogy and lessons learned. Thus, we hope readers find the analysis presented in this book helpful and inspiring to launch multiple initiatives that further accelerate the *global* deployment of electric vehicles.

July 2014

David Beeton
Gereon Meyer

Contents

Part I
Industry Perspectives

EV Business Models in a Wider Context: Balancing Change and Continuity in the Automotive Industry

Peter Wells and Paul Nieuwenhuis

Abstract This paper seeks to balance the tendency to analyze EV business models in isolation by setting them in a wider context in which the automotive industry is seeking to reconcile continuity and change in an increasingly volatile and uncertain competitive environment. This paper argues that one reason for the relative lack of penetration of EVs and the relative failure of the organizational innovations that have accompanied them is that there are powerful forces for continuity. It is argued that there are many other aspects of change in the industry that have been neglected by EV protagonists and policy-makers but which, for the industry itself, may be more urgent. Consequently, policy-makers may need to reflect upon the utility of traditional market incentives, research support and regulatory pressure.

Keywords Electric vehicles · Business models · State intervention · Automotive industry · Tesla · Autolib

1 Introduction

Is the Paris Autolib scheme a viable business model? How far do the innovations introduced by Tesla constitute a radical innovation in business model terms? How disruptive are innovations in EV business models for the existing automotive industry? This paper explores some of these and related themes by taking a wider, and longer, view of the automotive industry. In so doing, we urge some caution to those that wish to herald a new dawn in automobility, but equally recognize that in an industry facing myriad pressures there is an appetite for technological and

P. Wells (✉) · P. Nieuwenhuis
Centre for Automotive Industry Research, Cardiff Business School, Cardiff University, Cardiff CF10 3EU, UK
e-mail: wellspe@cardiff.ac.uk

P. Nieuwenhuis
e-mail: nieuwenhuis@cardiff.ac.uk

D. Beeton and G. Meyer (eds.), *Electric Vehicle Business Models*,
Lecture Notes in Mobility, DOI 10.1007/978-3-319-12244-1_1

organizational innovation, albeit tempered by considerable uncertainty as to what the future holds.

The paper commences with an account of the underlying logic behind the quest for an alternative to the mainstream automotive industry business model, propelled by the unique characteristics of EVs and the support 'ecosystems' they require [1]. Thereafter we highlight the ways in which there are enormous forces for path dependent inertia in the industry as a whole, and in wider society, that may act to ameliorate the incentives for business model innovation [2].

Moreover, as we subsequently argue, the issues surrounding EV technology and business model innovation are hardly the only pressures exerted on the contemporary automotive industry, and hence the issue of EV business model innovation needs to be placed in a broader strategic context. In particular, at a time of considerable economic austerity in many markets with the attendant financial pressures on revenues and profitability for vehicle manufacturers and their suppliers, the industry must simultaneously adopt a fundamental shift in global capacity locations and parallel development of several key technologies. Thus in the penultimate section of this paper we consider whether prevailing government interventions, both regulatory and fiscal, have been sufficient to compensate for the risks the industry faces both within the EV sector and more generally. We conclude with a call for stronger public-private partnerships that embody a long-term vision for the future of mobility in our societies.

2 Business Model Innovation and EVs: The Search for the Right Formula

The re-emergence of the EV as a potentially viable alternative to traditional petrol or diesel cars has been accompanied by an expectation that with it will come dramatic changes in 'automobility' and in the business models that vehicle manufacturers and others would need to access the market [3]. These twin expectations derive from the particular characteristics of EVs, particularly with regard to the initial stages of developing a market for these cars.

Compared with equivalent conventional cars, EVs have a high initial purchase price, more limited range, greater sensitivity to weather and other driving conditions, uncertain rates of depreciation, and lower expectations with regard to the longevity of the powertrain. While there are offsetting performance attributes that count to the advantage of EVs for consumers, such as smooth power delivery and quiet operation, the main financial attraction lies in reduced running costs (chiefly due to the lower cost of using electricity)—although such reduced running costs may still not entirely compensate for the initial higher purchase cost. In addition, EVs need access to a network of domestic, corporate and public recharging points. The concerns over the issue of range, and the time required for recharging, make the provision of accurate and timely information on infrastructure availability and road network conditions critical elements of the entire EV experience, and hence the wider EV 'ecosystem' in which business model innovation has been expected to occur includes electricity generation and distribution companies (including resellers and virtual aggregators), the manufacturers of recharging equipment, the installers of the recharging network, managers of that network, financial organizations to enable smart card payment systems, mapping and traffic information providers, public authorities at all spatial levels, software providers, and multiple organizations prepared to install recharging points [4, 5].

The orchestration of these multiple agencies is in itself a daunting task, and one of the key reasons why there has been an expectation of innovative business models, particularly from new entrants. The mainstream automotive industry, with its established business model and significant sunk costs, may be expected to show some reluctance in nurturing an alternative where it has the potential to render their existing investments redundant—a form of 'incumbents curse' [6]. Moreover, the low initial volumes expected for EVs opens up more opportunities for small and innovative companies to re-write the rules of competition.

Perhaps as a consequence of these considerations, the existing mainstream vehicle manufacturers have been relatively cautious in their pursuit of innovative business models, despite initiatives such as the Peugeot Mu concept or the Nissan-Sumitomo 'second life' idea to re-use automotive batteries in static emergency back-up operations. Rather, for the vehicle manufacturers, the preferred option may be to make EVs as much like traditional cars as possible by driving down battery and system costs and hence reducing purchase price or lease rates for consumers. As incumbents, the established vehicle manufacturers have some advantages that should not be discounted. These advantages include brand recognition, established retail and support networks; vast experience in marketing; vehicle integration competencies, political leverage; financial depth; and, profound technical skills. Furthermore, a cautionary approach has been somewhat vindicated by some high-profile failures in terms of business model innovation around EVs, most notably of course with regard to Better Place. As is explained further below, even the 'success' stories of business innovation, such as that of Tesla, deserve rather more careful evaluation when the industry is considered in historical perspective.

Governments at national and local level have also experienced some policy challenges with regard to EVs, particularly where the agenda has been to leverage investments in local recharging infrastructures in order to attract investment from the EV manufacturing sectors. Similarly, government support in terms of R&D programs and other incentives has some political attractions but is inevitably also fraught with hazard—as the media debate over A123 Systems and Fiskar (both recipients of US government funding and subsequently declared bankrupt) has shown.

As a consequence there is a sense in which there is an apparently lucrative market with high growth potential tantalizingly close, but which needs a judicious choice of intersecting business models at the right time, in the right place, and with the right combination of participants in order to unlock that opportunity.

Despite these comments, the runaway success in the EV sector has been the Renault-Nissan alliance, at least in terms of the sheer number of cars built and in use. The Nissan Leaf in particular does not embody any radical departure in terms of the business model, although Renault has sought to separate the battery from the car in terms of ownership and payment.

2.1 The Tesla Case

With the introduction of the Model S in mid-2013, Tesla became synonymous with entrepreneurial success in the nascent electric mobility market with strong product reviews for the car and optimism for the business model underwriting a surging stock market price in the US. In many regards the celebrations and plaudits appear somewhat premature. Bearing in mind the tumultuous history of EV pioneers such as Think (four times bankrupt in the 20 years to 2011), the financial position of Tesla does not appear particularly strong notwithstanding the backing of Elon Musk. As of the quarterly report filed on 9th August 2013, Tesla had grown revenue in the 6 months to June, 2013 to US$401 million, compared with just US $22 million in the 6 months to June, 2012. However, Tesla still reported a net loss (US$30 million), albeit reduced from the equivalent period in 2012 (US$105 million).

From Table 1 it is notable how far Smart and others have pre-figured many of the business model innovations offered by Tesla, including (with Daewoo) fixed price or 'no-haggle' sales. Smart embodied a high risk strategy for Mercedes as it simultaneously included a radical new vehicle design intended for a novel market segment, a new factory of unusual design and process operation, new supplier relationships, new distribution and sales including of course a new brand, mobility packages including tie-ins with rail service providers, and a focus on 'green' consumers. Many years of missed targets and accumulated losses followed. In 2013 Bernstein Research awarded Smart the dubious accolade of being the biggest loss-making model in the history of the European automotive industry, accumulating losses of £2.82 billion or £3,763/car [12]. Daewoo, another pioneer of alternative business model concepts albeit for different reasons (essentially to compensate for uncompetitive products), fared even less well—eventually the company was absorbed by GM.

Other vehicle manufacturers have struggled to make major shifts in their operational expertise or approach to market. Ford, in around 2000 under Jac Nasser, attempted to shift the centre of gravity of the business downstream by buying into, for example, car servicing and repair company KwikFit and vehicle dismantling operations. The project (termed Ford 2000) was rapidly abandoned. Others have sought to introduce new brands to access distinct market possibilities, again with largely unhappy outcomes. Mercedes failed with the luxury Maybach brand, just as GM did with the 'youth' Saturn brand. Alongside these notable failures from within the industry, there is a growing list of start-up failures or companies in tenuous

Table 1 A comparison of the business model innovations from Tesla with pre-existing innovations

Tesla innovation	Pre-existing example
Ownership of retail outlets	Very common and long-established practice in continental Europe; adopted by Daewoo as a market entry strategy in the UK in the 1990s
Creation of 'boutique' retail or experience outlets in shopping malls and other mixed retail locations	Attempted by Smart during early phase of marketing. Parallel examples include the Toyota Amlux Centre in Tokyo and the VW Wolfsburg 'Autostadt' brand experience facility
Fixed price, 'no haggle' retailing	Adopted by Daewoo as a market entry strategy in the UK in the 1990s
Provision of free access to unlimited charging via own fast-charger infrastructure (30 min recharge)	No comparable example, but many instances where new cars have been offered with 12 month supply of petrol
Battery swap system (on Model S) allows replacement in 90 s at US$60–80/swap	Initially tried by better place (now bankrupt). Only works for Tesla S models despite US $500,000 cost/swap station
Cars built to order, not sold 'off the lot'	Very common in Europe, particularly for prestige and sports cars for at least a proportion of total output. Morgan is a good example
Ordering new cars via retail outlet or internet	Internet retailing is well established, though because of legal constraints orders still need to be routed via dealerships
High levels of vertical integration (estimated at 70 % by value)	Historically common (e.g. Ford; VW); widespread for key technologies and materials e.g. fuel cells; magnesium
Introducing new brand, new model and new manufacturing facility simultaneously	An even more ambitious version of this was attempted at launch by Smart with the Hambach plant

Source [7–11]

positions in the nascent EV sector including Fisker, A123 Systems (Batteries), Coda Automotive, Bright Automotive, Aptera Motors, Miles Electric Vehicles, Ecotality (recharging networks), Next Autoworks Co. and of course Better Place (recharging networks).

With some 20,000 Model S cars planned for production it is by no means clear that Tesla can generate sufficient funds for expansion of the product range and the deployment of the recharging infrastructure. The business model, however, also benefits from the revenues Tesla obtains selling zero emission credits to others (at one stage amounting to 12 % of revenues), and from the revenues obtained from selling battery packs to Daimler (ironically, for the electric version of the Smart used in the innovative Car2Go schemes) and Toyota (for the PHEV version of the RAV4). Both Daimler and Toyota are investors in Tesla. Crucially, Tesla is not just a tale of entrepreneurial guile and fortitude: Tesla obtained a US$465 million government loan on extremely generous terms while the initial public offering in

2010 raised US$226 million. While Tesla repaid the loan almost 10 years early, in 2013, that initial vote of confidence helped underwrite the expansion of the business at the Freemont assembly plant in California. Some concerns have been raised about this sort of funding:

> "Personal loans made in 2008 by Elon Musk, Tesla's co-founder and CEO, provide a telling contrast. Musk received a much higher interest rate (10 percent) from Tesla and, more importantly, the option to convert his $38 million of debt into shares of Tesla stock. That's exactly what he ended up doing, and the resulting shares are now worth a whopping $1.4 billion—a 3,500 percent return on his investment. By contrast, the Department of Energy earned only $12 million in interest on its $465 million loan—a 2.6 percent return" [13].

This is a substantive issue for the future of government intervention. It is readily apparent that simple reliance on 'the market' to bring forth appropriate solutions is not plausible in many contexts. Indeed, there is much to the view that it is government or public-funded R&D (including that conducted at universities) that has generated most of the key technological breakthroughs of the contemporary era, from decoding the human genome to the creation of the Internet. Equally, government helps frame the market in important ways through fiscal regimes, regulatory interventions and the ability to underwrite long-term structural investments.

2.2 The Autolib Case

Despite the apparent differences, there are some similarities between Tesla and Autolib, the EV rental scheme pioneered in Paris. In the case of Autolib there is also the distinctive involvement of a wealthy individual, key investment and planning decisions from government, and partnership with the mainstream automotive industry. As with the Tesla case, the business logic behind Autolib is not entirely reducible to that of building and selling EVs.

Autolib is run by the Bolloré Group. While traditionally outside the automotive industry, a key interest for the Group is in lithium metal polymer (LMP) battery technology for which they hold all the patents. The battery division of Bolloré Group, Batscap has two battery factories: one in Quimper in Brittany; and a second in Montreal, Quebec. Initially, Bolloré approached car manufacturers in 2002 with their LMP technology but to no avail. In 2004 the company then partnered with CeComp in Italy, who developed the car, which in turn was designed and is built by Pininfarina as a subcontractor to CeComp, using Batscap batteries. The so-called 'Bluecar' has a 250 km range (urban cycle), 150 km on mixed cycle, four seats, and takes 8 h for a full charge (2 × 16 amp can reduce charging time from 8–4 h). The chassis is a combination of steel and aluminum, the body panels are aluminum, with some plastic panels (e.g. bumpers).

Bolloré won the Autolib tender in early 2011 with this vehicle, the Bluecar. It was able to combine the vehicle with its own in-house data management and automated interface terminal maker IER, which is a world leader in terminals for public services (e.g. automated check-in at airports). For Autolib IER supply the access card, charge points, kiosks (where you can sign up), operations centre, and information management. For Autolib, IER can keep track of cars through both GPS and via the charging points. Bolloré was also able to use Polyconseil—its own in-house telecoms consultant. Hence Bolloré had a neatly intersecting set of competencies to provide both the vehicle and the management of the system. These competencies make a difference, both to the overall value proposition and the success of the scheme. In particular subscribers (and those wanting to subscribe) can interact via a screen or via telephone to obtain help and information.

Autolib covers 47 towns in the region of Ile de France, with Paris at the centre. By mid-2012 Autolib had 1,740 Bluecars, 500 stations and 600 staff with a target of 3,000 cars, 1,000 stations and 1,200 staff by the end of 2013. By mid-2013 there were reportedly 82,000 subscriptions sold from late 2011 onward. Users can choose from three tiers of membership, with an additional cost depending on how much they drive. Autolib' memberships can last a day (€10), a week (€15), a month (€30), or a year (€144). About 35,000 are members with a yearly subscription. Once a subscriber joins, they are able to use the cars as often as desired subject to finding an available car and payment of the in-use fee (€7 per 30 min for day members, €6 for week and month-long members, and €5 for annual users).

Each Autolib recharging station has 4–6 spaces, and a terminal for signing in. Some 250 sites also have charging for other EVs (one space for a car and one space for an electric two-wheeler). The 47 municipalities pay €47,000/station as a subsidy, but Autolib pay a fee for the parking spaces, which will repay this subsidy by 2014—4 years ahead of plan. Once Autolib is profitable, profit will be shared with the municipalities.

Various claims are made about the impact of the Autolib scheme in terms of traditional vehicles replaced, CO_2 emissions reduced, and the contribution made connecting the city centre with the outlying suburbs. To date, however, no comprehensive evaluation has been undertaken. As a result, critics of the scheme are concerned that, for example, Autolib is primarily a substitute for public transport not private cars and hence does little to improve the environmental performance of the whole transport system.

Interviews with Autolib personnel in mid-2012 revealed that the average rental was 40 min and 10 km for Premium subscribers. Monthly subscribers tend to use the cars for longer each trip, typically about 3 h each rental. In addition, 70 % of Autolib users are in the 18–34 age-group: Contrary to expectations, tourists do not as yet constitute a high proportion of Autolib users.

Total investment in the whole project so far is €1.7 billion Euros, including cars, batteries, and infrastructure: Mostly from Bolloré. However, in turn Bolloré obtained a significant European Investment Bank loan of €75 million in 2012. Formal public investment is limited to the subsidy offered per charging station. Hence for the 'public purse' the scheme offers excellent value for money at relatively low risk. A crucial question in the context of this paper, however, is whether the project and the business model adopted is sensible for Bolloré. As a stand-alone activity there must be some considerable doubt over the viability of the Autolib project for Bolloré, at least in the short term. The number of vehicles is relatively low (only 3,000 projected) and once into a regular replacement cycle the Autolib 'market' might constitute 300 new vehicles per annum, but replacing these vehicles are an investment cost for Bolloré. Vandalism and accidental damage to the vehicles is an ongoing cost problem. Revenues from the subscription fees and use fees will of course become the most important income stream from the project but to date no figures have been released on this matter.

The business case for Bolloré begins to look rather more plausible when the wider context is considered [14–16]. In brief, Bolloré is looking to expand the business in a number of ways. First, private individuals can now lease the cars at €500/month, which includes a charging point. Second, Bolloré will now also sell the cars for €12,000 while renting the battery for €80/month. Third, the company is expanding the service side of its business with new markets in Lyon (Bluely) and Bordeaux (Bluecub), and most recently Indianapolis. Fourth, in a press release of the 12th September 2013 it was announced that Bolloré had signed a letter of intent with Renault for the joint development of car-sharing solutions and of new EVs, including the possible construction of a three-seat vehicle using Bolloré battery technology. With all these further developments, the Autolib case can be seen as a valuable shop window for Bolloré from which a much larger, longer-term and profitable business can be constructed around their intellectual capital and unique operational expertise.

3 Constraints on Innovation: Continuity in the Automotive Industry

It is worthwhile considering just why the established automotive industry is considered as relatively slow to change [17]. There are multiple facets to this issue, including both practical considerations such as the availability of appropriate skills and capacities, and more nebulous issues such as the lessons learned from historical experience.

Clearly, the vehicle manufacturers have an established workforce and physical assets premised on the design, integration, sourcing, manufacture, distribution, sale and support of all-steel vehicles with petrol or diesel engines. Such assets require large investments of substantial longevity. Moreover, the opportunities for switching may be constrained by external factors. For example, a wholesale shift into alternatives to the all-steel body is effectively impossible, because there is not a sufficient supply base for aluminum or carbon-fiber reinforced composites at present to substitute for the volumes accounted for by steel. Vehicle manufacturers are understandably reluctant when the supply of a key material or component is constrained, and therefore 'risk assessment' forms an important element in any strategic purchasing decision. It is telling that BMW felt the need to secure supplies of carbon fibre reinforced composites for the i3 and i8 models, for example.

This case is also an interesting illustration of the ways in which the reduction of life-cycle carbon emissions attributable to the use phase (currently circa 85 % for a standard car) throws increased attention on the carbon cost of manufacturing, and hence the need to mitigate emissions in this area.

Furthermore, incremental gains have been quite effective for the industry as a strategy to ameliorate some of the apparent advantages of EVs. In terms of environmental parameters therefore the advantages of EVs over conventional vehicles given an EU electricity generation mix are relatively modest [18, 19]. Regulatory regimes over carbon emissions have been largely shaped around a pace of change deemed acceptable—essentially political compromise with the industry—rather than that which is technically possible or environmentally imperative [20–22]. Incremental change therefore has been hard-wired into the industry by regulatory frameworks—although it could be argued that this position is changing. Moreover, fundamental assumptions about what constitutes a vehicle and how it may be used are also largely informed by the pre-existing industry, making it difficult for more radical concepts to establish an appropriate place in our pantheon of mobility opportunities.

4 Countervailing Pressures for Change in the Automotive Industry

The focus of attention on EVs is understandable, but for senior management in the vehicle manufacturers it is only one area of strategic concern demanding attention and resources. The following may be highlighted:

- Imbalanced capacity demanding closures in some locations, and new plants in others.
- Shortening product cycles, increased market fragmentation, and greater market volatility requiring a much larger product range.
- Requirements for a portfolio of new technologies of which EVs are but one.
- Shifting cultures of automobility in mature markets, along with ‘peak car’ saturation, resulting in concern for future revenue growth.

All of the above issues are significant for costs, revenues, or both. Some threaten to undermine the existing business model, which is interesting as a potential stimulus for the search for alternatives. However, contemporary economic conditions in the period since 2008 (especially in the EU) have not been conducive to growth, putting pressure on the ability of the industry to generate sufficient investment resources internally. In this context, the somewhat innate industry tendency towards conservatism is reinforced by necessary financial prudence in the face of increased volatility and uncertainty. What is most likely is the emergence of what might be termed ‘portfolio’ strategies whereby vehicle manufacturers seek to calculate a workable balance of product segment, technology and relative sales that both meets regulatory demands and still generates profitability.

Electric vehicles are of course part of this wider story. The pressure to develop EVs and new ways of bringing them to market is just one aspect of the multi-facetted challenges confronting the industry in terms of new technologies: existing petrol and diesel engine improvements; hydrogen fuel cells; hybrid systems; integration with mobile communications and mapping systems; new generation safety systems; and lightweight design via magnesium, aluminum and plastics are all individually demanding significant resources. Again, the industry has been here before. With the emerging technology of fuel cells, or with the strong possibility that magnesium would become a significant material in the future, vehicle manufacturers have sought strategic responses that enhance technical understanding and/or control over key resources. Typically, these responses have involved alliances, joint ventures or outright acquisitions as was the case with Daimler and Ford in the example of Ballard (fuel cell manufacturer). Whether such responses are enduring rather depends upon whether the technology or material does subsequently become of strategic significance, and whether the partner businesses can sustain some competencies that others may desire.

Fragmenting markets into a larger number of smaller product niches, combined with compressed model cycles, threatens to negate the benefits achieved via vehicle architecture strategies and overall result in a more turbulent market environment in which sales forecasts are increasingly difficult. Perhaps more significantly in the longer term are more nebulous concerns around an apparent shift in cultures of automobility that in the mature markets entails a downgrading of car ownership and use by younger people (the so-called 'peak car' phenomenon). The vehicle manufacturers, faced with the near certainty of saturation, can only increase revenues by capturing more value per vehicle: it is by no means self-evident that such a strategy can be realized by all. Overcapacity in the mature markets can only be resolved by expensive and socially-divisive plant closures, absorbing the resources and management attention that really needs to be concentrated on expansion in emerging markets.

5 Market Incentives or a New State-Business Relationship?

It is a debatable point whether EV registrations have been disappointing. The expectations of policy-makers and much of the media may not have been met as yet, but equally those expectations may have been unrealistic. The hopes of the industry have not been met either, but theirs is a more tempered concern. Early sales of hybrids, it is argued, were also modest. However, a stronger consensus appears likely over the need to orchestrate the novel powertrain and weight reduction strategies that must form the basis of the reinvention of the automotive industry. In most instances it is likely, again, that the vehicle manufacturers are central to this orchestration, bringing together constellations of companies that might hitherto have been largely outside the industry.

What is somewhat more problematic to forecast is the future participation of the state, which has been instrumental in seeking to establish the nascent market for EVs thus far [23–26]. The contemporary state intervention model does not stand up particularly well to detailed scrutiny either because of concerns over value for money against the risks taken, or because of competition between locations. Moreover, rules over state aid often preclude state involvement beyond the early stages of R&D, but often this means innovations are still well short of genuine market readiness. That is to say, there tends to be something of a 'desert' between prototype or demonstration levels, and the typical mass production scales associated with the high-volume automotive industry. Possibly initiatives like the 'Proving Factory' [27], established in the UK with the help of state funding to help bridge the prototype to mass production gap are one of the new ways in which state intervention can help partner with industry and thus bring products to the market.

Consumer or market incentives, on the other hand, do little more than send a message that this is a 'challenged' product that needs a sweetener before consumers will accept it. The use of such incentives is as flawed as those offered to scrap apparently perfectly functional cars after ten or even 8 years of use, justified at the time as a means of stimulating demand. Incentives distort the market for a modest short-term benefit and while in reality they do not 'cost' the governments in question a great deal, neither do they really help a great deal either. The short-term attitude inherent in much of the political process, wherein governments always have a wary eye on the next forthcoming election, makes policy announcements about incentives rather appealing regardless of their efficacy.

If there is continued market failure, which is certainly still an area to be debated rather than simply accepted, then the state needs to uncover better ways to resolve that failure. The Autolib case illustrates that the partnership approach whereby the state effectively underwrites the creation of a market space for EVs can work as an initial means of nurturing the transition to higher volume manufacture. The Tesla case in turn illustrates what can be done via entrepreneurialism, with the proviso that there is some concern over the future of standards and inter-operability with the Tesla model.

The notion of some form of collective rather than individual ownership that the Autolib case illustrates is probably a portent of things to come, particularly with the ongoing growth in various forms of car sharing schemes that both ensure a more efficient use of a given fleet of cars, and often speak to the operational advantages of EVs [28–30]. Again, car sharing is in the early stages of development and of course need not necessarily involve EVs, but the deployment of EVs via one or other version of such schemes is again an area of substantive potential; car sharing schemes may offer protected market niches, perhaps allied with 'committed' consumers that subscribe to the ambition of more sustainable mobility [31–34].

6 Conclusions

Business models in the EV world are still emerging and evolving, as they must do in response to shifting contexts. What we have sought to argue here is that the notion of a business model needs to extend well beyond the formal boundaries of the business itself; and beyond the boundaries of a particular point in time. Companies like Bolloré have the resources and the insight to aim at the long-term prize, so it would be a mistake to rush to judgment on an unfinished strategy.

On the other hand, perhaps the state also needs to be more innovative and to take more risks, but also to be more interested in sharing the rewards if those risks are then translated into success. In this sense, perhaps the state needs to act more like an investor, and in so doing can perhaps channel the rewards of investment back into further risk underwriting activities.

The business models literature tends to rather underplay the role of the state as a pivotal agency defining market possibilities, and yet a business model in a sector like that for EVs that ignores the role of the state is absolutely doomed to fail; indeed it makes little sense at all. The state at national and international level is pivotal in regulatory framing, and in steering all manner of developments that are critical to the success or failure of EV business models. Hence it is only through active engagement and interaction between the state and business that innovative business models will be able to deliver more sustainable mobility.

In the longer term the incremental extensions of EV business models either backwards or forwards along the value chain may come to be seen as temporary expedients. Alternatively, these developments and others like them may just be the start of a fundamental shift in the architecture of sustainable business as organizational forms come more closely to enabling the realization of government policies promoting the circular economy.

References

1. Roland Berger (2011) Automotive landscape 2025: opportunities and challenges ahead. http://www.rolandberger.com/expertise/industries/automotive/2011-02-28-rbsc-pub-Automotive_landscape_2025.html. Accessed 24 Mar 2011
2. Wells P, Nieuwenhuis P (2012) Transition failure: understanding continuity in the automotive industry. Technol Forecast Soc Chang 79:1681–1692
3. Abdelkafi N, Makhotin S, Posselt T (2013) Business model innovations for electric mobility-what can be learned from existing business model patterns? Int J Innov Manag 17(1). doi:10.1142/S136391961340003
4. Waller B (2011) Developing a new plug-in electric vehicle ecosystem for automotive distribution. Paper presented at the 19th international gerpisa colloquium on 'Is the second automobile revolution on the way?' 8th–10th June, Paris, 2011
5. Kley F, Lerch C, Dallinger D (2011) New business models for electric cars—a holistic approach. Energy Policy 39:3392–3403
6. Bock AJ, Opsahl T, George G, Gann DM (2012) The effects of culture and structure on strategic flexibility during business model innovation. J Manag Stud 49(2):279–305

7. Lewin T (2004) Smart thinking: the little car that made it big. Motorbooks, London
8. Winton N (2005) Investors to Mercedes—shut smart or fix it, January 30th 2005. http://www.wintonsworld.com/cars/carnews/carnews-2005/shut-mart-or-fix-it.html. Accessed 15 Sept 2013
9. Smith A (2006) Daewoo's U.K. factory stores: different, but not better, automotive news, September 25th 2006. http://www.autonews.com/article/20060925/SUB/60918046/daewoos-u.k.-factory-stores:-different-but-not-better#axzz2g00gyrYu. Accessed 15 Sept 2013
10. Marketing Week (2001) Daewoo's drive to survive in the UK, August 23rd 2001. http://www.marketingweek.co.uk/daewoos-drive-to-survive-in-the-uk/2042954.article. Accessed 15 Sept 2013
11. Godsmark G (1996) Daewoo sells cars … by buying them itself, The independent, August 24th 1996. http://www.independent.co.uk/news/business/daewoo-sells-cars–by-buying-them-itself-1311228.html. Accessed 15 Sept 2013
12. Walker S (2013) Financial disasters: the cars that lost a fortune, September 27th 2013. http://cars.uk.msn.com/features/financial-disasters-the-cars-that-lost-a-fortune#image=12. Accessed 8 Oct 2013
13. Woolley S (2013) Tesla is worse than Solyndra, May 29th 2013. http://www.slate.com/articles/business/moneybox/2013/05/tesla_is_worse_than_solyndra_how_the_u_s_government_bungled_its_investment.html. Accessed 4 Oct 2013
14. Ingram A (2013) French electric car-sharing service Autolib coming to … Indianapolis ?!?, , 11th June 2013. http://www.greencarreports.com/news/1084727_french-electric-car-sharing-service-autolib-coming-to-indianapolis. Accessed 11 June 2013
15. EVWorld (No date) Bolloré to sell bluecar for €12,000, rent battery for €80/Month, http://evworld.com/news.cfm?newsid=29834. Accessed 15 Sept 2013
16. Davies A (2013) We tried out the French electric car-sharing service that's on its way to the US, July 22nd 2013. http://www.businessinsider.com/reviewing-paris-autolib-car-share-system-2013-7?op=1. Accessed 15 Sept 2013
17. Steinhilber S, Wells P, Thankappan S (2013) Socio-technical inertia: understanding the barriers to electric vehicles. Energy Policy. doi:10.1016/j.enpol.2013.04.076
18. Granovskii M, Dincer I, Rosen MA (2006) Economic and environmental comparison of hybrid, electric and hydrogen fuel cell vehicles. J Power Sources 159:1186–1193
19. Hawkins TR, Singh B, Majeau-Bettez G, Strømman AH (2013) Comparative environmental life cycle assessment of conventional and electric vehicles. J Ind Ecol 17(1):53–64
20. Euractiv (2013) MEPs back fuel-efficiency plan for cars, including 'super-credits', April 25th 2013. http://www.euractiv.com/energy-efficiency/meps-back-fuel-efficiency-plan-c-news-519335. Accessed 18 Sept 2013
21. ACEA (2013) Long-term CO_2 targets must be ambitious and scientifically-founded, not 'political', April 23rd 2013. http://www.acea.be/news/news_detail/press_release_long-term_co2_targets_must_be_ambitious_and_scientifically-fo. Accessed 20 Sept 2013
22. Cars21 High Level Group final report on the competitiveness and sustainable growth of the automotive industry in the European Union, 2012. http://ec.europa.eu/enterprise/sectors/automotive/files/cars-21-final-report-2012_en.pdf. Accessed 20 Sept 2013
23. Villareal A (2011) The social construction of the market for electric cars in France: politics coming to the aid of economics. Int J Automot Technol Manag 11(4):326–339
24. Nair R, Miller-Hooks E, Hampshire RC, Bušić A (2012) Large-scale vehicle sharing systems: analysis of Vélib'. Int J Sustain Transp 7(1):85–106
25. Brand C, Anable J, Tran M (2013) Accelerating the transformation to a low carbon passenger transport system: The role of car purchase taxes, feebates, road taxes and scrappage incentives in the UK. Transp Res Part A: Policy Pract 49:132–148
26. Regulation (EC) No. 443/2009 of the European Parliament and of the Council of 23 April 2009 setting emission performance standards for new passenger cars as part of the Community's integrated approach to reduce CO_2 emissions from light-duty vehicle, Official J Eur Union, 5.6.2009
27. See http://www.theprovingfactory.com/

28. Harmer C, Cairns S (2011) Carplus annual survey of car clubs, transport research laboratory (TRL). Published Project Report PPR567. http://www.carplusorguk/wordpress/wp-content/uploads/2011/07/Carplus-Annual-Survey-of-Car-Club-Members-201011-Finalpdf
29. Truffer B (2003) User-led innovation processes: the development of professional car sharing by environmentally concerned citizens. Innovation 16(2):139–154
30. Glotz-Richter M (2008) CarSharing as part of transport integration concepts: experience of the Bremen mobil.punkt. Public Transp Int 57(2):30–32
31. Cerfontaine C (2011) Combined mobility: or offering citizens the possibility to live in city without owning a car. Public Transp Int 60(2):28–30
32. Bieszczat A, Schwieterman J (2012) My car, your car ⋯: more and more people are driving a communal automobile. Planning 78(5):37–40
33. Firnkorn J, Müller M (2012) Selling mobility instead of cars: new business strategies of automakers and the impact on private vehicle holding. Bus Strategy Environ 21(4):264–280
34. Cepolina EM, Farina A (2012) A new shared vehicle system for urban areas. Transp Res Part C: Emerg Technol 21(1):230–243

Four Business Models for a Fast Commercialization of Plug-in Cars

Mats Williander and Camilla Stålstad

Abstract Plug-in vehicles are one important means to lower CO_2 emissions from the transport sector. Despite this, uptake is slow. This can be well explained by theory on social dilemma problems and on diffusion of innovations. The traditional "sell-and-disengage" business model is not suitable for plug-in cars. Using an entrepreneurial business model generation process we have developed four alternative business models that address important factors for the speed of which customers adopt an innovation. The results show that alternative business models are necessary, but they cannot alone ensure a fast, and lasting, commercialization of plug-in cars. As a complement, governments will have to take measures to control external factors that influence the viability of business models for plug-in cars.

Keywords Plug-in car · All-electric car · PHEV · REV · Business model · Social dilemma problem · Diffusion · Innovation

1 Introduction

Plug-in vehicles are seen as one important means to lower CO_2 emissions from the transport sector and to reach a fossil independent vehicle fleet by 2030, a goal set by the Swedish government.[1] The slow commercialization of plug-in vehicles seems to have taken people by surprise.[2] What is surprising though are the high expectations

[1] See http://www.regeringen.se/sb/d/15703/a/196433 (read 2013-09-03).
[2] Illustrations: http://www.theengineer.co.uk/blog/uk-electric-car-take-up-is-slow-while-carmakers-charge-ahead/1016123.article or http://www.bloomberg.com/news/2013-05-27/german-automakers-falter-in-meeting-million-electric-goal.html or http://www.gizmodo.com.au/2012/01/electric-cars-off-to-a-slow-start/ (read 2013-09-03).

M. Williander (✉) · C. Stålstad
Viktoria Swedish ICT, Lindholmspiren 3A, Gothenburg, Sweden
e-mail: mats.williander@viktoria.se

C. Stålstad
e-mail: camilla.stalstad@viktoria.se

D. Beeton and G. Meyer (eds.), *Electric Vehicle Business Models*,
Lecture Notes in Mobility, DOI 10.1007/978-3-319-12244-1_2

of a rapid shift to plug-in cars, and even more surprising is the stubbornness most plug-in carmakers show when sticking to the same business model they use for traditional cars, despite the disappointing sales figures achieved when it is used for plug-in cars. The slow uptake of plug-in cars can be well explained by theory on social dilemma problems and on diffusion of innovations. By building on these two theories, we have developed four alternative business models that address the reasons why the traditional car sale business model is inefficient for plug-in cars. The business models' viability have been tested, and from that, conclusions can be drawn about specific societal support that will be required to avoid the risk of future social dilemma problems that may obstruct a fast introduction of plug-in cars.

The paper is structured as follows. First we set the terminology on plug-in cars used for this chapter. Then a brief description of the business model concept is given. After that we explain why the current business model for cars does not work for plug-in cars, after which we look into what issues a business model for plug-in cars should address. Thereafter follows a description of the business model generation process we have used for developing the four business models, which are then depicted through a common template. We then revisit the social dilemma problem, where governmental intervention is most likely required for a diffusion to take place and then not discontinue. The chapter ends with a concluding section including a description of limitations.

2 Terminology

The terminology for plug-in electric vehicles is not yet well defined. The study behind this chapter has focused on cars, although our findings may be relevant also for other vehicles. With plug-in cars we mean all cars that can be charged from the grid, i.e. both *all-electric* cars, like Nissan LEAF and the Tesla cars, plug-in hybrid (*PHEV*) cars, like Volvo V60 PHEV and Toyota Prius Plug-in, and range extender cars (*REV*), like Opel Ampera and BMW i3 with the range extender option. Since all plug-in cars, except the all-electric cars, can easily replace a traditional internal combustion engine (ICE) car, we have had the all-electric car in focus when designing the alternative business models. The business models should however be applicable, potentially with some adjustments, also to the other types of plug-in cars.

3 What Is a Business Model?

A business model is a description of how a company creates, delivers and captures value [1, 2]. A viable business model must provide value to the customer that is higher than the costs for providing it, and then capture the difference. Business models unleash technologies' inherent value with different degrees of efficiency and with different characteristics [3]. In the most common "sell and disengage" business model, the ownership of a product, and all its future costs are transferred to the buyer in exchange of money at point of sale, possibly with some warranties included. An alternative business model can be to keep ownership of the product and sell access to it as a subscription, like renting an apartment.

It is the business model that defines what the offering will be compared with. A traditional "sell-and-disengage" business model for plug-in cars will make the established ICE cars the obvious reference point, while a carsharing service like Car2Go can have taxi and public transport as reference points.

A key to a viable business is the business model owner's ability to capture the excess value, i.e. the value above the cost for providing it. Since there is limited willingness to pay privately for something that will benefit the commons [4] this is extraordinary challenging when a substantial part of a product's value is common good and not private good. This will be further discussed in the section on social dilemma problems.

4 Why the Current Business Model for Cars Doesn't Work for Plug-in Cars

The "sell-and-disengage" business model where the physical product ownership, and hence all risk, is transferred to the buyer at point of sale may work fine for established technologies with low perceived risk, but not for novel technologies, especially not when they are not up to par with the technology they challenge. Using the traditional "sell-and-disengage" business model for plug-in cars encourages customers to use the ICE car as reference, hence giving three issues unnecessary focus; (1) the higher price, (2) the shorter driving range, and (3) the uncertain battery life length. The hesitance this creates among new car buyers becomes also valid for used car buyers, which makes the used car value uncertain.

People use a higher implicit discount rate for technologies that are unfamiliar to them [5], i.e. they demand a lower "price per utility" than from the established technology. Humans also have a nonlinear perception of gains and losses, where we perceive the punishment from losing as bigger than the joy from winning an equal amount of value [6]. This means that when a new technology performs worse than the established technology in any attribute, it will easily be discarded as inferior.

Price versus operating cost also matters. We can learn from behavioral economics that a high initial price but low future operating cost often is perceived as less attractive than a lower initial cost but higher operating cost, even when the total economic impact is exactly the same [6, 7]. Renault's decision to sell their all-electric cars without the battery and instead sign up the customer on a forcing monthly lease is an example of change in the temporal distribution of financing to make the all-electric car offering price- and cost-wise more similar to prices and costs of ICE cars.

5 Issues to Address for a Successful Diffusion of Plug-in Cars

Innovations are, by definition, new and unfamiliar at the time they start to diffuse among adopters, and this goes for plug-in cars as well. An innovation is an improvement compared to an existing state or practice among the intended adopters. Plug-in cars, and especially the all-electric car, are an illustration of a valuable solution to a problem that is of common responsibility. The private interest to pay is however dependent on the private good the car provides, and this can create a so-called social dilemma problem. Hence, two general issues to consider when trying to commercialize plug-in cars are how innovations are diffused successfully and how to handle social dilemma problems.

5.1 Social Dilemma Problems

A social dilemma problem is when "individuals in interdependent situations face choices in which the maximization of short-term self-interest yields outcomes leaving all participants worse off than feasible alternatives" [8]. The value a technology provides can be split into common value and private value. The value can be positive, like improved mobility, or negative, like noise, accidents and emissions.

Plug-in cars provide more common value (or more precisely, less negative common value) than ICE cars, for instance less noise, local air pollution and tailpipe emissions. On the other hand, they have difficulties in providing the same amount of private value as the ICE car in terms of for instance price, driving range, charging(/refueling) time, and towing capacity, although they often manage to show a lower energy cost. Hence, the slow adoption of plug-in cars can be seen as an illustration of a social dilemma problem. When most car buyers continue to choose the ICE car because of its perceived higher short-term private value, we may all be worse off.[3]

5.2 Diffusion of Innovations

The relative speed with which intended adopters, like potential car customers adopt an innovation is highly dependent on five factors [9, 10]. There are of course more factors that affect the adoption speed, but these five are considered most important:

[3] c.f. the Stern Review on the Economics of Climate Change, executive summary: http://webarchive.nationalarchives.gov.uk/20130129110402/http://www.hm-treasury.gov.uk/d/CLOSED_SHORT_executive_summary.pdf.

(1) *Its relative advantage* in terms of economy, initial cost, comfort, social prestige, time and effort saving, and immediacy of rewards. Here, plug-in cars hardly provide clear benefits. Its costs relative the ICE technology are uncertain. Comfort and prestige may be similar to those of ICE cars while time and effort savings depend on its use. It can be better in commuting situations where charging takes place at home and/or at work and hence is more effortless than refueling at a gas station, but it is worse at long-distance driving, even when fast-charging.
(2) *Its compatibility* with sociocultural values and beliefs, with the technology it is compared against, and with potential customers' need for the innovation. Here, plug-in cars fit well with contemporary values and beliefs in Sweden. The compatibility with ICE cars is insufficient. All-electric cars cannot replace ICE cars to 100 %. Many car owners/users are quite happy with their ICE cars and don't see the plug-in car as delivering innovations they personally have been missing.
(3) *Its complexity*, i.e. how difficult it is perceived to understand and use. Here, plug-in cars *can* be easy to use and understand, but are not on par with ICE cars, for instance in providing understandable, reliable and consistent driving range predictions, which are crucial to support range anxious drivers.
(4) *Its trialability*, i.e. if it can be tested on a limited basis. Many auto dealers do not even have plug-in cars in their showrooms, and those who have, offer the same level of testing as for ICE cars. This means you can test drive the car for a while, but not really evaluate it to see if the car really fits in your daily life.
(5) *Its observability*, i.e. its visibility to others. Many plug-in cars are variants of ICE cars, which make them difficult to distinguish from their more common car cousins. This makes them relatively unobservable in the streets, which in addition makes them difficult for potential owners to use as identity markers.

In summary, plug-in cars have had difficulties in providing sufficiently convincing personal advantages for potential customers. However, the business models used for diffusing plug-in cars do not seem to have addressed the above five factors in order to achieve a successful diffusion.

6 The Business Model Generation Process

A look at the Swedish plug-in car market revealed an astonishing lack of alternatives to the traditional sell-and-disengage business model. In order to appraise alternative business models, which were lacking in reality, a project was set up where the objective was to create four alternative business models and validate them in accordance with recommendations from successful serial entrepreneurs. It was decided to follow an entrepreneurial Customer Discovery business model generation procedure [11–14] where an initial business model hypothesis evolves through a refinement procedure with tests against the assumed customer segment, financial calculations, and so forth. The hypothesis is refined in cycles until there is an offer that actually is sold to members of the final customer segment.

The alternative business models to be developed should assume plug-in cars maintain the attributes they have today and hence focus on use-cases and customer segments in which the unfavorable attributes of plug-in cars mattered less and the favorable attributes could be emphasized. The business models should also consider the most important five factors for successful diffusion of innovations and try to deal with the social dilemma problem of many plug-in cars.

Initially, 17 business model hypotheses were generated, discussed, compared, combined and briefly checked, either against potential customers or in discussions with companies in similar businesses. They were then reduced in a combination-selection process, inspired by the Pugh Concept Selection Method [15] until four business models remained.

The four business models have not been validated as far as to actual sales since the project was without commercialization intent. However, each business model was refined to the point where potential customers showed substantial interest in the value propositions at prices and costs that were judged as realistic by the companies in similar businesses we used as reviewers of the business models. Reviewers were typically car rental companies, car leasing companies and carsharing companies. The customers that the business models were tested against were first chosen according to each business model's customer segment hypothesis, and then found in our network of friends and friends' friends in the Gothenburg region in Sweden. Interviewers were selected so that no interviewer had previously met or talked to the interviewee.

7 Business Model Descriptions

The business models will be described using a specific structure. Each description starts with *the initial idea* where the initial hypothesis is briefly depicted. *How it works* describe the final business model's function. *Diffusion strengths* highlight the most important strengths among the previously described five diffusion factors, relative to the sell-and-disengage business model. Finally, *viability factors* describe the top most important factors we identified that economically break or make the business case for the business model.

7.1 Conditions

The business models have been developed with all-electric cars in mind, although they might be of value also for plug-in hybrid electric cars and range extenders. Calculations for all four business models (which are available upon request) are

Table 1 Reference cars

Car attribute	Nissan Leaf	VW Golf 1.6 TDI BMT
Electricity consumption (kWh/km)	0.173	n/a
Diesel consumption (l/10 km)	n/a	0.38
Tailpipe emissions (CO_2 g/km)	0	99

based on an interest rate of 4 %, a fuel price of €1.59/l petrol or diesel, and an electricity price of €0.1225/kWh.[4]

We have used the Nissan Leaf and a Volkswagen Golf as reference cars in business models BM1, BM3 and BM4, their attributes are specified in Table 1. A Volvo V60 DrivE Momentum and a broader range of plug-in cars are used as references in BM2 due to a significant difference in customer segment preferences.

Additional business model information such as descriptive folders, storyboards, and short illustrative videos can be found at http://www.viktoria.se/projects/believe.

8 Four Alternative Business Models

Here follows a description of each of the four business models that were created by the project using the business model generation process.

8.1 BM1: Free-Floating All-Electric City Cars

8.1.1 The Initial Idea

The inspiration for this business model was the fact that more than half of the world population now lives in cities and that it will be more or less impossible for cities of the future to carry the load of today's car density per citizen. Hence, the initial idea was to provide all-electric car based service, providing personal mobility as flexible and as private as a personal car, but with more convenience and less hassle than private car ownership in a city. The idea was also to offer a range of all-electric cars, from vans to Tesla Roadster, so that almost any mobility need or desire could be met.

8.1.2 How It Works

You go by taxi but you are the driver. That is how free floating carsharing works. As user, you pay per minute of use, and that's it. The typical customer is a city dweller who finds it increasingly annoying to own a car in the city but still wants

[4] These are valid amounts as of June 2013 in Sweden at an exchange rate of 9.10 SEK per €.

personal mobility beyond what public transport and taxis can provide. With free-floating all-electric city cars, you don't have to pre-book, stick to a certain time interval or leave the car where you took it. It's as a taxi, but you drive yourself.

A free-floating carsharing service is established within a designated city zone. Within this area, users may pick up and return the cars at any public parking spot. The user pays per minute of use. Electric energy, maintenance, road tolls etc. are included. The customers are believed to be both private persons and employees on business errands. Typical private customers are people who have no car or families who use the service instead of owning a second car. It is a flexible complement to public transport, just as a taxi is. For people living in city centers it is an added benefit not having to bother about parking. Companies may also benefit from using the all-electric carsharing service instead of compensating employees for driving their own cars and having parking places for them.

Customers book and have contact with the service through a smartphone, tablet or computer, showing where cars are located. The idea is that there should always be enough cars available for the users to mostly book just before use, i.e. the cars are not supposed to be reserved until shortly before they are used. In this business model, it is assumed that the city contributes with free parking if the cars are all-electric. Firstly because an all-electric car doesn't contribute to local air pollution and secondly because a carsharing car replaces 9–13 privately owned cars [16] and hence the service can help reduce car density per citizen. The combination of free-floating carsharing service with all-electric cars is hence superior to a similar ICE solution both for society and for users.

8.1.3 Diffusion Strengths

This business model completely avoids the ICE car as reference point and any car attribute comparison, since it provides a service that rather competes with taxi and public transport. Relative to them, this service costs more than public transport but is more flexible, private and convenient while it costs less than taxi but requires the customer to drive and park.

The strengths relative to the sell-and-disengage business model are:

Relative advantage: It provides access rather than ownership, and 2–3 times higher car utilization than privately owned cars.

Compatibility with sociocultural values and beliefs: It supports the general trend where the interest in owning a car is decreasing [17] and where people move to cities.

Complexity/simplicity: Given that one has a smartphone, which is required to use the service and locate the cars, the service is very simple.

Trialability: The service can be tried out easily and without any long-time commitment. One only pays for the time the car is used.

Observability: It is the visibility of the service rather than the cars' make and model that matters here. Striped[5] cars at the required density within a designated city zone will ensure a continuous high visibility within that zone.

8.1.4 Viability Factors

A high utilization of these cars is the main tool to make a service like this better with all-electric cars than with ICE cars. The lower mileage cost will offset the higher price.

However, the current battery warranty conditions show to be the main financial constraint since we have to expect a car's residual value to be almost zero when the battery warranty ends.

The second most important factor for the profitability of this business model is that the city can provide low-cost or free parking for electric cars.

A drawback of this business model is that it requires many cars at launch. Customers will only be satisfied if they can easily find a car within a reasonable distance, typically about 200–300 m, and that the designated area is not too small. This can easily sum up to a need of several hundred cars, i.e. a significant investment and hence business risk.

8.2 BM2: Plug-in Cars as Company Cars

8.2.1 The Initial Idea

Company cars, cars provided by the employer for employees to use as their own private cars, are common in Sweden. About 25 % of all new car sales in Sweden are company cars used for this purpose.[6] Hence, by making plug-in cars attractive as company cars a significant part of the Swedish new car market will be available.

Company car holders are taxed based on the value of the car. For plug-in cars taxation is calculated on a reduced value of the car, making it advantageous for the employee to choose a plug-in car as a company car.

8.2.2 How It Works

The setup for company cars in Sweden ranges from financing models where the employer pays for all costs of the car and the employee is just taxed for the benefit, to financing models where the employee pays for all costs of the car with a gross salary deduction.

---

[5] Made unique in appearance with for example color and/or sticker film, see for example http://www.carscoops.com/2008/04/volvo-launches-new-personalized-sticker.html (read 2013-10-11).

[6] http://www.tjanstebilsfakta.se/artiklar/nyheter/?page=article&nid=827 (read 2013-10-10).

A growing number of companies offer company cars of the latter model, which is cost-neutral to the company. For the employee the cost for such a company car is somewhat lower than if he or she would have bought the same car privately.

For plug-in cars the financing model that is cost-neutral for the company is the worst-case scenario for the employee. The higher price of the plug-in car has to be covered by the employee through a higher gross salary deduction, which in turn needs to be offset by the lower taxation for the car and the lower mileage cost in order for the plug-in car to be competitive relative to an ICE car. Our calculations show that plug-in cars mostly cost the employee less per month than a comparable ICE car. This cost difference increases with increased driving.

In cases where the employer takes all or some of the costs for the company car a plug-in car can be an expensive alternative for the company. Normally, only the employee benefits from the lower tax and the lower mileage costs of the plug-in car, while the employer is affected by the higher purchase price. To make the plug-in car an attractive alternative to both parties, the idea is that the employer, through an additional gross salary deduction gets a share of the benefits which compensates for the higher price of the plug-in car. Even though the full benefits of the plug-in car in this case do not reach the employee, the plug-in car can still be a competitive alternative to an ICE car.

Company cars in Sweden are often used as the family's first car. The car is hence expected to be able to take the whole family and luggage on holiday trips etc. Many company car holders also use their car a lot for business travelling, driving long distances. For these reasons all-electric cars might not be the primary choice as a company car. However, if the employer offers, for example, a car swapping service among colleagues, for use when the all-electric car is insufficient, some of these difficulties might be overcome.

8.2.3 Diffusion Strengths

The main strength of this business model is solidly shown in the fact that such a substantial share of all new car sales in Sweden are company cars. Some diffusion issues are however worth discussing:

Relative advantage: Compared to other company cars, both PHEV and REV cars show to be less costly while they also are 100 % replacements of ICE cars. This should put them on many employees' consideration list if they only are aware of their existence.

Compatibility with sociocultural values and beliefs: PHEV and REV offer the customer a solution that is as convenient as, less costly than, and more environmentally-friendly than a comparable ICE car, which is in line with current Swedish values and beliefs.

Complexity/simplicity: No major difference to private car ownership.

Trialability: This is the weak spot compared to many leasing schemes. This lease is on 60 months, which may be perceived as a significant commitment compared to

the 36 months used for many ICE cars. 60 months is required to get a sufficiently long depreciation time and still have a lower monthly cost. The alternative would be a high used-car market risk, which most likely would make the business case unprofitable for the customer.
Observability: No major difference to private car ownership.

8.2.4 Viability Factors

The most important viability factor for this business model is the Swedish tax regulations for company cars. The reduction of taxation for plug-in cars is however only temporary, and for this business model to be viable it has to be extended, and in the future even adjusted to compensate for the expected decrease of difference between running costs for plug-in cars and ICE cars.

The second most important viability factor is the value of the car on the used-car market. A higher used-car value would further improve the business case of plug-in company cars and/or enable a shorter (more competitive) leasing period.

8.3 BM3: All-Electric Car Subscription

8.3.1 The Initial Idea

This business model is an attempt to increase the utilization degree of cars owned by private persons. The calculus for the all-electric car requires high utilization of the car for it to be competitive compared to a corresponding ICE car. The average privately owned car in Sweden is driven only 11,820 km/year,[7] corresponding to 32 km/day, which is not even a quarter of the maximum range of a normal all-electric car. Based on these figures we saw great potential in utilizing the cars more, hereby achieving a more competitive cost structure for the all-electric car. The initial idea was inspired by consumer-to-consumer carsharing and when combined with the social trend of decreasing interest in car ownership the result became a carsharing service suitable for car commuters.

8.3.2 How It Works

The all-electric car subscription is a carsharing service for frequent users, people who need access to a car more or less every day. For these frequent users today's carsharing services become too expensive to compete with car ownership. The subscription offers access to an all-electric car at an agreed level, at a fixed monthly

7 See http://trafa.se/PageDocuments/2012.xls (read 2013-09-10).

fee corresponding to the cost of owning a corresponding ICE car. If the subscriber needs to use a car more than agreed in the contract, excess usage will be charged corresponding to fees paid by regular carsharing customers.

The all-electric car subscription allows the subscriber to use an all-electric car for commuting. When the commuter does not use the car it will be available for regular carsharing customers, hence the utilization of the cars is maximized. When using the all-electric car for commuting it is possible to keep the car at home during the night, and as long as it is fully charged in the morning the night hours are free of charge. The subscription includes all costs of the car, even electricity (fuel). The subscriber does not have to worry about services, reparations and maintenance of the car.

All-electric car subscribers also have full access to the regular carsharing service. This gives the subscribers easy access to a conventional car when the all-electric car is not sufficient, for example for longer trips. The subscriber pays the ordinary carsharing service fee for using a conventional car, but has no costs for the subscription during the time and hence no double car costs occur. All car booking is easily handled through an application available for smartphones.

A positive effect of having subscribers commuting by carsharing cars is that they at the same time are moving cars to where the demand for car access is. During the days the cars will be parked in areas where many people work and during the evenings and weekends the subscribers bring the cars to suburbs and areas where many people live. This movement of cars helps extending the market for the carsharing service.

8.3.3 Diffusion Strengths

This business model lowers the threshold for trying an all-electric car for commuting. Since there is no investment to be done, and the subscription does not imply any long-term commitment the risk that the customer takes by trying this service is minimal. By having the customers pay only for their actual usage of the car, the cost is also competitive, compared to owning a car used for the same purpose.

The strengths relative to the sell-and-disengage business model are:

Relative advantage: It provides access rather than ownership, and 2–3 times higher car utilization than privately owned cars.
Compatibility with sociocultural values and beliefs: The business model supports the general trend where the interest in owning a car is decreasing [17]. Driving an all-electric car is also more environmental-friendly than driving a comparable ICE car, which is in line with current Swedish values and beliefs.
Complexity/simplicity: The service is very simple and flexible, hence adapts to the customers' needs. A fixed all-inclusive monthly fee makes it easy for the customers to predict their car costs.

Trialability: The service can be tested without any long-time commitment, and no admission fee is required. The subscription also makes it possible for carsharing operators to include all-electric cars in their range of cars offered to regular carsharing customers since high utilization of the all-electric cars is necessary to achieve viability. Offering all-electric cars to regular carsharing customers makes it possible for more people to get experience from all-electric cars.
Observability: The visibility of the service could be optimized by using striped[5] cars and by branding the service as a smart environmental-friendly alternative to owning a car.

8.3.4 Viability Factors

Occupancy rate of the cars is the most critical factor for the viability of this business model. The low operating cost and high purchase price of all-electric cars means that the more the car is used, the better the viability. The usage is however limited by the battery warranty. Since the car is so highly utilized the mileage limit of the warranty is reached rather soon. This severely limits the earning opportunities of the business. If the battery warranty was extended, or if it was concluded that the all-electric cars will be functional even after the warranty has expired, it would have a very positive impact on the viability of this business model.

This business model is easiest to realize as a complement to an existing public carsharing service. This way the existing customer base creates demand for the cars during periods when the subscribers are not using them.

8.4 BM4: Leasing Chain for All-Electric Cars

8.4.1 The Initial Idea

Rental car companies in Sweden annually buy more than 20,000 new cars. These cars are used in the business for about 18 months before they are sold on the used-car market. If plug-in cars could take a substantial share of this flow they would help establish a used-car value, which is important for a rapid commercialization of plug-in cars.

It might however, be tricky to create both enough demand for plug-in cars and to achieve a profitable business case for the plug-in cars on the rental market, especially for all-electric cars. This has given birth to the idea of the leasing chain in which the all-electric car is kept until its end of life, hereby eliminating the residual value issue.

This business model was initially focusing on finding a way to use all-electric cars in car rental services, but in the resulting version the first customer in the lease chain might as well be a private household, a carsharing service (see BM1 or BM3), a company car provider (see BM2) as a car rental company.

8.4.2 How It Works

Some claim that all-electric cars already today provide lower total cost of ownership (TCO) than comparable ICE cars for many consumers [18], but this is only the case if seen over the total lifetime of the car. If the car is sold before that, the TCO will highly depend on the used car price when sold. Today, there is a considerable uncertainty about used car prices for plug-in cars. One way to reduce this uncertainty is to own the car until its end of life. Most private households are uninterested in such commitments. In fact, a growing share of households is not interested in owning a car at all, as long as they have access to one [17].

The idea with this business model is to let a leasing company own the car and lease it out to a chain of customers until its end-of-life. The potentially lower TCO can then be shared between the leasing company and its customers, and the residual value risk is significantly reduced. Since all-electric cars are expected to need fewer repairs when getting older compared to ICE cars it may be that only all-electric cars can be considered for a used-car operational lease offer. If so, that can be a significant and lasting advantage for all-electric cars in the lease market.

The leasing chain offers operational lease of all-electric cars in a chain spanning several customers. When the car is new it can be leased by car rental companies, carsharing companies and other new-car leasing customers at about the same price and on the same terms as comparable new ICE cars. In the second, third and possibly fourth leasing scheme, the typical customer segment is two-car households in suburban areas who commute daily by car. The reason to focus on households with two cars is that the ICE car can be used in cases where the all-electric car is not sufficient.

The older the car, the more economical an all-electric car is compared to an ICE car. Through all the leasing chain the value proposition must be compared with ICE lease on a per-km basis since the financial lease cost will be higher for an all-electric car but together with the lower running cost becomes much more comparable, even lower.

8.4.3 Diffusion Strengths

Relative advantage: Operational lease gives customers peace of mind, especially compared to owning an old ICE car. A leasing offer for every car age preference is novel and wanted in Sweden, especially offers for leasing of cars with an age that matches households' view of the second car.

Compatibility with sociocultural values and beliefs: Many suburban car commuters are well aware of their carbon footprint but have difficulties in finding reasonable alternatives to car commuting in their "cash-rich but time-poor" lives. Leasing of an all-electric car could offer a competitive solution to this dilemma.

Complexity/simplicity: All costs and leasing time are pre-known. An all-electric car can often be perceived as simpler to use than an ICE car since it doesn't have to be refueled.

Trialability: The business model provides opportunities to compare how an all-electric car fits with one's lifestyle with a commitment limited to the shortest offered leasing period.
Observability: No major difference to private car ownership.

8.4.4 Viability Factors

The top viability factor in this business model is the distance limitation in the battery warranty. Car commuters typically drive more than 20,000 km/year, which quickly accumulate to distances beyond what's warranted.

The second most important factor is the battery warranty time. Many customers are willing to use cars that are older than what is currently covered by the warranty.

The third most important factor is that the difference in running costs between the all-electric car and an ICE car must cover the price difference. As ICE cars become more fuel-efficient, this cost difference may erode and hence destroys the business case.

9 Social Dilemma Problems Revisited

An economic analysis of the four described business models shows that their viabilities are dependent on four recurring factors, namely:

- The battery warranty conditions
- The energy cost gap per driven km
- The price gap between a plug-in car and a comparable ICE car
- The technology improvement speed.

If we examine these factors further, it can be seen that the causes to these factors are not business model related but rather technology related. The battery warranty conditions set the limit for the accumulated distance that can be driven at low risk; this combined with the lower energy cost per km for plug-in cars can define how much of the price gap can be recovered during use. The technology improvement speed affects the depreciation rate of sold plug-in cars. The speed of technology improvement is often something wanted, while the strategy among customers to wait to purchase because improved technology is around the corner is unwanted, as it delays the diffusion of the technology wanted by society.

The complex network of technologies behind these four factors is constantly evolving. How the factors develop relative to each other will therefore have a major impact on the size of the social dilemma problem of plug-in cars, and hence have a major impact on the diffusion rate of plug-in cars in society. There is a considerable risk that over time these factors will increase rather than decrease the social dilemma problem for plug-in cars, but there are steps that governments can take to mitigate these risks:

- The price gap may remain or widen: There is a fierce competition and significant overcapacity in the car industry today while plug-in cars are sold in low volumes, which combined may widen the price gap. In addition, governmental incentives to plug-in car buyers may end. One mitigation option can be a bonus-malus system where buyers of ICE cars with high CO_2 emissions have to pay fees, which then are used as rebates to buyers of less CO_2 emitting cars. Such a bonus-malus system can be designed to be cost neutral for the government and hence be long-lasting.
- ICE cars may become more fuel efficient, not least because of EU's ambition to legislate further CO_2 tailpipe emission reductions for cars. One mitigation option is to increase petrol- and diesel taxes accordingly so that the energy cost gap between ICE and all-electric cars remain or increase.
- The battery warranty/battery life length might not improve: Our analysis suggests that the battery warranty time/distance is more important than driving range. Customer segments can always be found where the current driving range is more than sufficient while the battery warranty time directly affects the depreciation and hence the economic competitiveness across most business models and customer segments. Battery warranties vary between countries and states. Nissan gives a battery warranty of 8 years or 160,000 km in the US while they give 5 years and 100,000 km in Sweden, which seems to be without reason. One way to guide interest into the battery warranty issue can be for governments to legislate minimum warranties, as many governments have done for consumer goods. For instance legislate that the best available global battery warranty must count also in this country.
- The technology speed may induce a waiting strategy among potential customers. Some governments and municipalities, with Norway as a periphrastic example, gives from time to time various forms of incentives to plug-in car buyers, like lower tax, free parking, driving in bus lanes, exemption from congestion charges and so forth. These incentives can not last forever, especially not when the market share of those cars become significant. One way to mitigate the depreciation caused by technology improvement speed can be to let incentives follow the car for its lifetime. By that, older plug-in cars may be valued higher than otherwise if they are accompanied by highly valued incentives.

10 Conclusion

As has been shown here, alternatives to the most common "sell-and-disengage" business model can be designed to provide a more successful diffusion of plug-in cars. This can be done by designing plug-in car based offerings that give the addressed customer segment a reference these offerings can compete with, and then use available knowledge on how to successfully diffuse innovations when crafting the business model around the offering and its customer segment.

The history of business provides a wealth of business models that have been tested in various businesses and business conditions over the years, while research on entrepreneurs and start-up companies gives a structure for how to validate business models at low cost [11–14, 19]. Combined, they constitute a useful toolbox for anyone who wants to design business models that better fit the potential customer segments for plug-in cars than the sell-and-disengage business model does.

Car manufacturers hold significant business model experience and innovativeness. Despite that, many of them go for the same business model for plug-in cars as for ICE cars. Tesla Motors however, who subsists on making only all-electric cars, show a much higher willingness to develop their business model in accordance to the business model principles and theories that have proven successful for introduction of new technologies, which is rewarded by the market through impressive sales figures.

Plug-in cars provide a lot of social value (common good) for which there is limited private interest to pay (limited private good). Without governance of the four technology-dependent factors we have identified, the diffusion of plug-in cars may not only be weak and delayed, but may also halt once it has started. Governments themselves may be an initiator, for instance through commendable efforts to reduce tailpipe CO_2 emissions from ICE cars.

10.1 Limitations

The four business models that have been described in this chapter have been designed with Sweden in mind, i.e. Swedish tax rules, taxes, prices and costs, the Swedish mentality among potential customers, Swedish commuting distances and so forth. This may not apply to other countries without adjustments or changes.

The business models, although validated to a certain extent, have not been validated to the level where customers actually buy. A considerable remaining business risk should hence be expected in each of the business models.

The cost and price calculations behind each business model have been made with rigor, have been crosschecked and have been discussed with companies in related businesses. However, the business model designers are not professional economists and it can not be expected that external parties would identify all possible flaws or reveal all flaws they possibly would identify.

Despite these limitations, we believe that some interesting analytical generalizations can be made.

Acknowledgments This chapter describes results from the Business model innovation for Electric Vehicles (BeliEVe) project. The BeliEVe project has been carried out by Viktoria Swedish ICT in cooperation with Ericsson AB, and with financing from The Swedish Energy Agency.

References

1. Björkdahl J (2009) Technology cross-fertilization and the business model: the case of integrating ICTs in mechanical engineering products. Res Policy 39:1468–1477
2. Osterwalder A, Pigneur Y (2010) Business model generation: a handbook for visionaries, game changers, and challengers. Wiley, Hoboken
3. Chesbrough H, Rosenbloom R (2002) The role of the business model in capturing value from innovation: evidence from Xerox Corporation's technology spin-off companies. Ind Corp Change 11:529–555
4. Linder M (2013) Capturing value from green offers—an examination of environmental differentiation and economic performance. Chalmers University of Technology, Department of Technology Management and Economics, Gothenburg
5. Geiler H, Attalie S (2005) The experience with energy efficiency policies and programmes in iea countries—learning from the critics. IEA Information Paper, Paris
6. Kahneman D, Tversky A (1979) Prospect theory: an analysis of decision under risk. Econometrica 47:263–291
7. Ariely D (2010) Predictably irrational, revised and expanded edition: the hidden forces that shape our decisions. HarperCollins, New York
8. Ostrom E (1998) A behavioral approach to the rational choice theory of collective action. Am Polit Sci Rev 92:1
9. Moore G (2002) Crossing the Chasm: marketing and selling disruptive products to mainstream customers, Revised edn. HarperCollins, New York
10. Rogers E (2003) Diffusion of innovations, 5th edn. Free Press, New York
11. Blank S (2006) The four steps to the epiphany—successful strategies for products that win. Lulu.com, Pascadero
12. Blank S, Dorf B (2012) The startup owner's manual—the step-by-step guide for building a great company. K&S Ranch Press, Pescadero
13. Furr N, Ahlstrom P (2011) Nail it then scale it: the entrepreneur's guide to creating and managing breakthrough innovation. Nathan Furr and Paul Ahlstrom, Provo
14. Ries E (2011) The lean startup: how constant innovation creates radically successful businesses. Viking, London
15. Frey D, Herder P, Wijnia Y, Subrahmanian E, Katsikopoulos K, Clausing D (2009) The Pugh controlled convergence method: model-based evaluation and implications for design theory. Res Eng Design 20:41–58
16. Martin E, Shaheen S (2011) The impact fo carsharing on household vehicle ownership. ACCESS: The Magazine of UCTC, University of California, Berkeley
17. Kalmbach R, Bernhart W, Kleimann P, Hoffmann M (2011) Automotive landscape 2025: opportunities and challenges ahead. Roland Berger Strategy Consultants, Munich
18. Werber M, Fischer M, Schwartz P (2009) Batteries: lower cost than gasoline? Energy Policy 37:2465–2468
19. Sarasvathy S, Venkataraman S (2011) Entrepreneurship as method: open questions for an entrepreneurial future. Entrepreneurship Theory Pract 35:113–135

Electrification of the Powertrain in Automotive Applications: "Technology Push" or "Market Pull"?

Vincent R.H. Lorentz, Martin M. Wenger, Reiner John and Martin März

Abstract Full battery electric vehicles are yet to achieve significant worldwide success on the market. This analysis shows that the required technologies have already been developed, but not for a use in the mass market, where low cost is mandatory to be successful. Central roles for this success will be played by governments, industries, and research and standardization institutions. A great effort in both national and international synchronization and coordination activities, together with a clear regulatory push, will be mandatory. Enabling technologies for the plug-in hybrid and full electric vehicles will also come from the "3Cs": Costs, Comfort, and Climatic dependency. There is definitely not a single impulse that will be sufficient to enable the market for electric vehicles.

Keywords Market push · Market pull · Electrification · Automotive powertrain · Plug-in hybrid and full electric vehicles · Battery systems · Fuel cells · Power electronics · Disruptive technologies

V.R.H. Lorentz (✉) · M.M. Wenger · M. März
Fraunhofer IISB, Schottkystrasse 10, 91058 Erlangen, Germany
e-mail: vincent.lorentz@iisb.fraunhofer.de

M.M. Wenger
e-mail: martin.wenger@iisb.fraunhofer.de

M. März
e-mail: martin.maerz@iisb.fraunhofer.de

R. John
Infineon Technologies AG, Am Campeon 1-12,
85579 Neubiberg, Germany
e-mail: reiner.john@infineon.com

D. Beeton and G. Meyer (eds.), *Electric Vehicle Business Models*,
Lecture Notes in Mobility, DOI 10.1007/978-3-319-12244-1_3

1 Introduction

1.1 Problem and Motivation

The powertrain of a road vehicle regroups all the components needed to transform the energy stored in chemical form (e.g., fuel, battery) into mechanical energy used for the propulsion. The term powertrain regroups the engine, the gearbox, the transmission, the driveshaft and the differential. A conceptual view of a hybridized powertrain for automotive applications is shown in Fig. 1. The unidirectional powertrain is the common powertrain based on a fuel or gas energy source which cannot be recharged by energy recuperation during the breaking phases. The bidirectional powertrain uses an electric energy storage unit that can be recharged through energy recuperation during the braking phases or through another path providing a source of electric energy linked to the unidirectional powertrain (e.g., an electric generator driven mechanically by an internal combustion engine). The first energy source represents fuel or gas, while the second energy source represents an electric energy storage system like batteries or super-capacitors. The first energy converter can be an internal combustion engine or a fuel cell with an electric motor. The second energy converter is an electric motor used as motor during the acceleration phases and as a generator during the breaking phases, thus recharging the second energy source. The load represents the power transmitted to the road.

Depending on the level of hybridization, the unidirectional or bidirectional powertrain are removed. In a pure internal combustion engine vehicle, only the unidirectional powertrain is implemented. On the contrary, in a pure battery electric vehicle, only the bidirectional powertrain is implemented. In hybrid vehicles, both unidirectional and bidirectional powertrains are present and during driving, both can also be active simultaneously, for example when the second energy source has to be recharged through the unidirectional powertrain during driving.

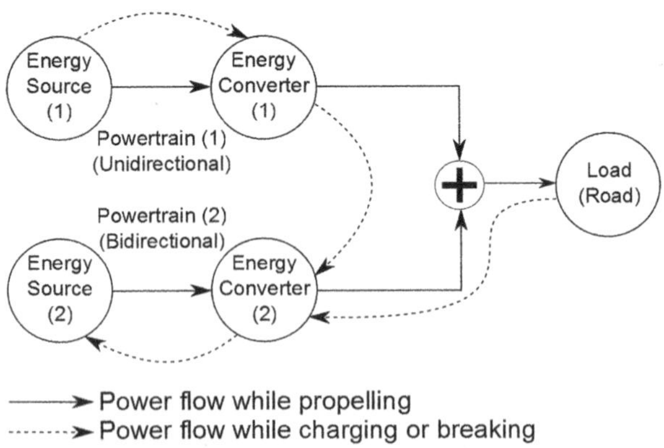

Fig. 1 Conceptual illustration of a hybridized powertrain [1]

The ongoing powertrain electrification in the automotive industry is motivated by several factors, but underlies some apparent paradoxes. The German government initiated a target for a million electric vehicles on the German roads by 2020 and subsidizes this by funded research projects of one billion Euros to develop technologies for intensifying the electrification of the powertrain. The electrification of the powertrain in automotive applications has been identified as a key strategic element for the future of the automotive market all around the world. The market run-up for electric vehicles was planned to begin in 2011; some 3 years later the market share for electric vehicles still remains extremely low.

1.2 Automotive Industry's Current Situation and Future

Clayton Christensen performed a very interesting analysis on the future of the automotive industry in 2003 in [2]. He analyzed that the ongoing transformation will result in a massive transfer of the ability to make attractive profits in the future away from the automobile manufacturers and toward certain of their suppliers. He has estimated this transformation to take one or two decades to complete.

By looking at the French car manufacturers (e.g., Renault, PSA Peugeot Citroën) and some of their biggest direct French suppliers (e.g., Valéo, Faurecia, Plastic Omnium), the analysis done by Clayton Christensen is mainly right today [3]. More and more competencies have been transferred to the subsystem suppliers and both the in-house production depth and the core competencies of the French car manufacturers have been decreased. Also the financial health of the French car manufacturers is currently not very secure.

The situation in Germany is quite different. Almost all German car manufacturers have focused on the top end of the market or have moved upmarket during the last 15 years (e.g., Volkswagen, Audi, BMW, Mercedes, Porsche). Opel has attempted but failed to move upmarket, exactly like Renault and PSA Peugeot Citroën. The new entrants coming mainly from Asia were successful in occupying the bottom of the market. The German direct suppliers (e.g., Bosch, Continental, ZF-Friedrichshafen) have mainly seen their competencies being extended, even if the hybrid branch is still today not the mainstream (i.e., where the money is made).

By pursuing the analysis from Clayton Christensen further, it can be observed today that the car manufacturers are aware that the money in the mobility business will not be only in the assembly of the subsystems. Also the role played by the brand of the car manufacturers is threatened. BMW for example is well known for its very well performing gasoline and diesel engines. This is a differentiating factor for the brand. In the future, if BMW assembles electric vehicles with the battery system coming from a supplier and the electric motor from another supplier, the BMW brand will lose a part of its flavor. The car manufacturers are aware of this, and in the case of BMW, they are for example developing their own solution to design battery systems. However, also the tier one (e.g., Bosch) and battery suppliers (e.g., Panasonic) are investing in the development of battery systems.

Today, for electric vehicles, it is still unclear from whom in the supply chain the battery system will come from (i.e., car manufacturer, direct supplier, battery cell manufacturer).

2 Barriers in the Powertrain Electrification

In the past, beginning in 1828, the electric vehicle has encountered a lot of technological issues. Unfortunately, most of these issues still remain very present today. The technological challenges are not only linked to the infrastructure but also to the powertrain, which means the electric motor, the battery and the electronics (i.e., power and control electronics). In 1828, the Hungarian inventor Ányos Jedlik invented an early type of electric motor and built a small car powered by his new motor. The invention of rechargeable batteries allowing electricity storage on board a vehicle happened in 1856 by the French physicist Gaston Planté. In 1911, the first gasoline-electric hybrid car was released by the Woods Motor Vehicle Company of Chicago in the USA. In 1902, the invention of the mercury arc rectifier used to convert alternating current (AC) into direct current (DC) by Peter Cooper Hewitt marked the start of the power electronics era.

2.1 Lack of Infrastructure

Concerning the infrastructure, electrification began in the late 1910s, but still during the 1920s, only few subscribers could be registered. In Europe, the real democratization of electrical power began in the 1950s. Looking at this, it becomes clear that the democratization of the electric vehicle does not only rely on a single technology, like the battery capacity, as it is often claimed. The democratization of the electric vehicle is dependent on many technologies that must optimally play together, thus forming a coherent whole.

The currently available infrastructure for charging electric vehicles is not very developed. The rarity of electrical charging stations as well as their localization is not encouraging the customers to buy electric vehicles. Furthermore, the long charging times are often perceived as a barrier. This is not a problem for most plug-in hybrid vehicles which are only rarely connected to the grid.

Before installing the necessary infrastructure, a common standard for the charger interface is required. Today, no global standard has been agreed worldwide. Adding to this the possibility to choose between AC and DC charging, there is still a long way to go for a worldwide common standard, assuming that one day there will be one.

An interesting parallel with the standardization of the charging connectors for smartphones can be drawn. In its press release IP/09/1049 entitled "Commission welcomes industry's commitment to provide a common charger for mobile phones" from June 29th, 2009, the European Commission has published the following:

> Incompatibility of chargers for mobile phones is a major inconvenience for users and also leads to unnecessary waste. Therefore, the Commission has requested industry to come forward with a voluntary commitment to solve this problem so as to avoid legislation. As a result major producers of mobile phones have agreed to harmonise chargers in the EU. In a Memorandum of Understanding (MoU), which was submitted to the Commission today, the industry commits to provide chargers compatibility on the basis of the Micro-USB connector. In addition new EU standards to ensure continued safe charger use will be developed to facilitate the implementation of the MoU. The first generation of new inter-chargeable mobile phones should reach the EU market from 2010 onwards.

Although compliance is voluntary, a majority of the mobile phone manufacturers have agreed to make their mobile phones compatible with the Micro-USB connectors. It is important to notice that the initiative was taken by the European Commission and was not coming from the industry nor from single governments. Another important fact that can be noted is that this standard has nearly become a worldwide standard now, even though it was initiated in Europe. Independently from the fact that now the customer does not have to worry about the specific type of charger needed when buying one, this standardization had a further significant impact. The so called *autonomy anxiety* (i.e., *range anxiety* in the case of electric vehicles) was minimized, as the customer knows that compatible chargers are easy to find, to get, to buy and to use. This has dramatically reduced the inconvenience resulting from the limited autonomy of modern powerful smartphones and tablets, and has boosted the sales of such mobile devices, even if the exact impact is difficult to quantify precisely.

2.2 Poor Performances of the Battery

The other weakness that is very often pointed out in electrified powertrains is the battery. It is generally stated that its energy capacity is too small and its cyclic and calendric lifetimes too limited. A part of this observation is related to the fact that the customers have had some negative experience with lithium-ion batteries from the consumer market, like batteries for laptops, smartphones or digital cameras. Most of these batteries do not provide an acceptable storage capacity after 3 years of heavy duty. However, what is mostly ignored by the customer (since the manufacturers are not granting access to such batteries in an easy way) is, that very high quality and performance batteries are available, especially for special applications, like for example in the military, in the aeronautic and in the aerospace domains. For example, the French company Saft provides lithium-ion batteries with proven performance for satellites with the specifications compliant with the NASA and ESA standards: long cycle-life (i.e., 5 years ground storage and up to 20 years in orbit) and long cycling capability (i.e., up to 100,000 cycles using adapted depth of discharge).

The message is clear: Most of the technological solutions for batteries are already available today, some of them even in Europe, but unfortunately, they currently have higher costs compared to the costs of available gasoline and diesel powered vehicles. A crucial point will be to reduce these costs and optimize the costs of the technologies so that they can be produced in high volumes at low costs. However, in Europe, it seems that the battery business is in a critical situation. No lithium-ion battery manufacturer in Europe is capable of producing advanced automotive battery cells and battery systems for the large car manufacturers at a competitive cost level. Europe appears to be divided on the subject and each member state tries to build up its own business, like for example Li-Tec with Daimler and Evonik in Germany, Bolloré Group with batScap in France, European Batteries in Finland. None of these projects can be considered a success today, and in consideration of the current activity level of the Asian battery manufacturers (i.e., Japan, Korea, China), thinking that a single member state in Europe can compete against Asia and the US will remain just a sweet dream. The only chance for the European battery business will come from putting the European efforts together. Unfortunately, time is running out, and with it, the challenge for catching up is increasing. In history, Europe has shown that it is able to be successful when the member states decide to work together, like what was done with EADS and Airbus.

2.3 Eternal Comeback of the Fuel Cell

Fuel cells convert chemical energy coming from a fuel into electricity through a chemical reaction with oxygen. The most commonly used fuel is hydrogen, but hydrocarbons (e.g., natural gas) and alcohols (e.g., methanol) can also be used. Like rechargeable batteries, the chemical reaction occurring in fuel cells is in theory reversible and called electrolysis, but this only works in specially developed reversible fuel cells currently only available in research laboratories. However, unlike batteries, fuel cells require a constant source of fuel and oxygen. The oxygen is in general taken from the ambient air. This is the source of one of the major issues of fuel cells, since the ambient air transports a lot of impurities that will contaminate the membrane over time. The membrane in fuel cells is a separating layer that acts as an electric charge exchanger (i.e., an ion exchanger) as well as a barrier film separating the gases in the anode and cathode compartments, comparable to the role of an electrolyte in batteries. In the past, several attempts were made, for example by Daimler, to introduce fuel cell vehicles to the market [4]. All of these attempts have inexorably failed, mainly because of five reasons [5]:

- High costs of the fuel cell technology (e.g., use of platinum required)
- Limited life time due to the contamination of the membrane
- High costs of hydrogen fuel (e.g., electrolysis as a process for converting electricity into hydrogen only has an efficiency of 75 %)
- Need for entirely new fuelling facilities and transportation concepts (e.g., available infrastructures not adapted to handle hydrogen)
- Competition from other technologies in the market (e.g., batteries).

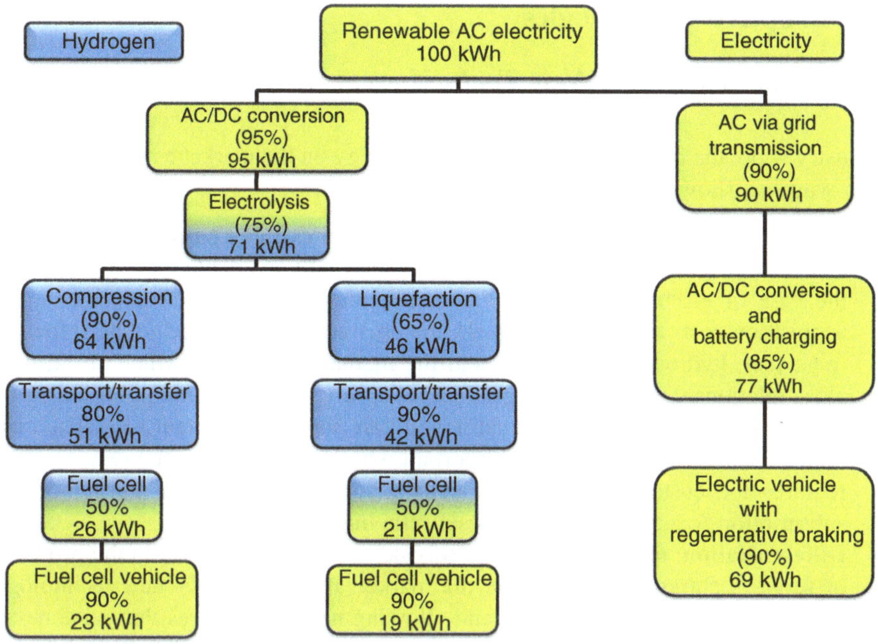

Fig. 2 Useful transport energy derived from renewable electricity [6]

Fuel cells are bringing together the drawbacks of oil and batteries. The fuel needs to be produced and transported in pipelines or on the road, like gasoline and diesel, but with hydrogen, there are some additional issues due to the storage pressure required or transportation requirements, thus adding losses to the transport process of hydrogen, as shown in Fig. 2. Further, the still unresolved problem of the contamination of the membrane is closely linked to the limitation of other promising technologies, like lithium-air batteries, that could boost the current energy storage density available in lithium-ion batteries by a factor of 3–5 in practice. The advantage of electricity is that there is no need of pumps nor tanks on trucks for transportation: Electricity simply flows through the already available infrastructure, thus making it in combination with batteries a much better candidate for powering electric vehicles, even if the existing infrastructure is insufficient and must be greatly improved and enhanced.

For supplying mobile applications like electric vehicles, electricity obtained from hydrogen fuel cells appears to be four times as expensive as electricity drawn from the electrical transmission grid used to charge batteries [6]. The hydrogen economy will definitely never make sense for electric vehicles in a mass market, but in niche markets and in the case of special applications, hydrogen energy should be considered. For example, in the case of vehicles with long driving ranges, fuel cells can be theoretically superior to batteries in terms of mass, volume and refueling time [7].

3 Disruptive Technologies

3.1 Introduction of a New Technology

Industry wide, the introduction of a new technology on the market can only occur in two ways, as shown in Table 1 [8]:

- **Overall cost leadership**: the new product or service can be provided at lower costs compared to the products or services currently available on the market. In the case of electric vehicles, a distinction must be made between the initial acquisition costs of the vehicle itself, the operating costs related to the refueling (e.g., fuel, hydrogen, electricity) and the maintenance costs, which are also in close relations to the provided product reliability. An additional point has to be considered because in some situations, it can play an important role: The customer's felt costs. This point is important when the product or the service often requires the customer to pay extra taxes actively, not offered as a monthly subscription (e.g., gasoline and diesel refueling costs).
- **Differentiation**: the new product or service cannot be compared to the products or services currently available on the market. It is an entirely new technology that is opening up new horizons and allowing new experiences that were never previously done or possible. An example of such a differentiation could be a product with enhanced valuable functionalities miniaturized like never done before (e.g., the smartphone). Smartphones have replaced most of the standard mobile phones today, but their batteries have 7 times less autonomy (i.e., only for 1 or 2 days), they cost 3 or 4 times more, but they actually provide a lot of very valuable functionalities for the customer: They are worth their price. In the field of electric vehicles, such a differentiation could be the function of true autonomous charging, which means that the electric vehicle will search for an inductive charging station autonomously (i.e., without any driver) and charge its battery on its own. This is not possible with gasoline or diesel vehicles without changing the entire refueling infrastructure. For the customer, autonomous charging means that the electrified vehicle will completely take care of the whole refueling process (i.e., recharging the batteries). For the recharging technology, this means that cable connections will have to be avoided and wireless charging will be the solution.

Table 1 Porter's generic market strategies [8]

	Uniqueness	Low-cost position
Industry-wide	Differentiation	Overall cost leadership
Particular segment only	Focus	

Table 2 Differentiation between technology push and market pull [10]

Description	Technology push	Market pull
Technology uncertainty	High	Low
R&D expenses	High	Low
R&D duration	Long	Short
R&D customer integration	Difficult	Easy
Time-to-market	Uncertain/unknown	Certain/known
Sales market-related uncertainty	High	Low
Kinds of market research	Qualitative-discovering	Quantitative-verifying
Need for change of customer behavior	Extensive	Minimal

3.2 Market Pull Versus Technology Push

Innovation stimuli commonly occur in two ways [9]:

- **Market pull**: In this case, the source of innovations is currently not satisfying the customer needs, thus resulting in needs and demands for new solutions. The concrete stimulus can come from single persons or a group of persons in the population.
- **Technology push**: In this case, the stimulus for the development of new products and processes comes from the research and development side. The followed goal is to make money on the basis of new technologies. In this case, it does not matter if a market demand currently exists or not.

As shown in Table 2, it can be differentiated between radical innovation (i.e., technology push) and incremental innovation (i.e., market pull) [9]. This means on one hand that technology push can be considered as creative or destructive and provide major improvements. On the other hand, market pull means the development of replacements or substitutes with small improvements. This description can be completed by the consideration that technology is particularly relevant for the early stages of the product life cycle, and that market factors are especially relevant for their further diffusion.

The deficiencies and shortcomings of technology push and market pull strategies are shown in Table 3 [9]. However, the differences between technology-induced

Table 3 Summary of deficiencies of technology push and market pull [13]

Technology push	Market pull
Risk of starting with what can be researched and evaluated easily	Risk of looking only at needs that are easily identified but with minor potential
Risk of addressing the needs of atypical user	Continuing to change the definition of the "opportunity"; "miss the opportunity"
Potential for getting locked into one technical solution	Lack of being a "champion" or "true believer"

approaches and market-induced approaches are not always as clear. It was observed that successful products are using a weighted combination of market pull and technology push approaches (i.e., push-pull), thus increasing the innovation degree of the company [11, 12]. A key for achieving this balance inside a company is the establishment of a precise internal communication between the technology oriented divisions (e.g., research and development, engineering) and the market oriented divisions (e.g., marketing, sales).

In the case of electric vehicles, based on Table 2, it can be concluded that a clear technology push is needed to adapt and provide existing cutting-edge technologies to the customer at much lower costs than what can currently be provided on the market. The potential customer must become familiar with these technologies and be convinced of their quality and reliability, so that a massive market penetration of electric vehicles can be enabled.

3.3 Regulatory Push

Several governments worldwide have tried policies to overcome existing barriers to the electrification of the powertrain in automotive applications. They have promoted the sales of electric vehicles and funded further developments like better battery technologies and other components. They have also established incentives to lower the purchase prices of electric vehicles. Some governments (e.g., France) are giving incentives on bought electric vehicles to accelerate the market penetration of these types of vehicles. Unfortunately, the charging infrastructure is still extremely weak, when not simply nonexistent, thus being a major obstacle for the potential customers interested in electric vehicles. However, the catalytic and synchronization roles of the policy makers are central and crucial.

In [15], an analysis of eco-innovations enabling the distinction between different environmental areas such as the reduction of energy use, and the pollution of air, water or soil, was done. The analysis shows that currently, most of the relevant eco-innovations having a high impact on the environment are focusing on energy saving, CO_2 emission reduction and recycling. In fact, policy makers are in general not interested in eco-innovations, but in specific environmental areas such as energy efficiency and renewable energies. However, in reality, cost reduction is the main goal of the reduction of the energy consumption, not pollution.

In Europe, the regulatory push is given by the European Commission, by the governments of the member states internationally and nationally, but also at some more localized levels, like for example what is happening in the Torino e-District, where local companies are working together to promote electric mobility locally and then more globally by developing an open hardware manufacturing platform for a small battery-EV. However, care must be taken when involving regulatory push with the goal to accelerate the adoption of a new technology. Through incentives, the market numbers are getting biased, and even uncompetitive technology can gain a certain amount of market share. Analyzing such markets and building a long-term strategy on this analysis is very dangerous.

Regulatory push should not occur by deciding about the technologies that must be used in the future. A clear and stable long-term pollutant emission reduction roadmap, definitely fixed on an international or at least continental level, would be key to a successful regulatory push concerning the reduction of emissions in road vehicles. Without forcing any specific technological solution, but rather forcing the reduction of the allowed pollutant emission levels, it is likely that full electrification of the powertrain would naturally emerge.

3.4 Market for the Electric Vehicles

Electric vehicles can be identified as a disruptive technology [16]. Searching for a market for electric vehicles requires strategies to find legitimate and unsubsidized lead adopters. Since electric vehicles do not currently satisfy the requirements of performance on the mainstream gasoline and diesel market, these vehicles cannot compete in this market. It is currently not clear where the market for electric vehicle is and will be, but on the other hand, it is also certain that the electric vehicle market is not in an established automobile market segment. Looking back 20 years, it appears that at several times, different brands of car makers all around the world have tried to focus precisely on the mainstream market with their developments of electric vehicles, mainly due to the reason that new or emerging markets do not provide enough profit to ensure the growing rate needed by big companies. In summary, this means that other ways must be found to the customers, thus implying that a market on which electric vehicles can be used must be found.

New or emerging markets are interesting because early entrants into disruptive technology markets develop capabilities that constitute strong advantages over later entrants [16]. In contrary, holding back from the market and waiting until research organizations have developed a breakthrough technology (e.g., in batteries, electric drives, lightweight materials, power electronics) is the path of least resistance. However, this strategy is in general not a route to success in the case of disruptive technologies, since the summed delays represent a handicap that will never be compensated again.

A further crucial point that must be clear is that a market that does not exist cannot be analyzed. In the case of the electric vehicle, market research cannot give an answer to what the early markets for electric vehicles will be. Customers cannot tell how they will use electric vehicles, because they will discover how they might use the product when they will have it. Real useful market information can only be gained through expedition through market analysis when the product is used by customers.

In such a market, the business plan to address a new market must be based on a clearning strategy and not on the execution of a preconceived strategy [16]. It is very probable that a better direction will appear after the first products have entered the market. It must therefore be planned that the initial plan will be wrong and be learned as fast as possible what will be right. It is therefore very important to conserve free resources to make it right on the second or third try.

4 Key Enablers for the Future Mass Market: The "3C"

The stepwise electrification of the powertrain and auxiliaries in road vehicles is an obligation. It is currently becoming a key competence in the worldwide competition of the automotive industry to deliver both low emission vehicles and increased performances and functionalities. The customers are requesting vehicles offering the same level of functionality and comfort at a nonincreasing cost level. In fact, the three situations that must be overcome for the electric vehicles to get a higher market acceptance are Comfort reduction, Climatic dependency, and Cost increase.

Today, none of these situations can be avoided and tradeoffs must be made when looking for vehicles with higher levels of electrification. To solve these issues, a holistic approach for the combined management of mechanical, electrical and thermal energies and their interactions on the vehicle level is required, rather than their independent optimization at component and subsystem levels. This holistic consideration will provide a competitive advantage to the automotive companies having understood this.

4.1 No Loss of Comfort

The comfort in a vehicle is mostly measured in a very subjective way. For example, range anxiety because of a too short maximum driving range and inaccurate visualization of the remaining energy in the battery of pure electric vehicles is perceived by the customer as a loss of comfort. Another example is the time needed to recharge the battery, perceived as much too long when compared to refueling an internal combustion engine vehicle. These weaknesses can and must be overcome by technology and promotion together. The customer's habits will be modified. Through inductive fast charging in combination with autonomous driving it will be possible to no longer think about recharging the batteries again and let the electric vehicle do it by itself. The driving itself can be made very comfortable since the electric motor develops its torque even at 0 rpm. Electric vehicles offer a vast new range of possibilities that have still to be discovered.

4.2 Low Climatic Dependency

Most of the customers think that with pure battery electric vehicles, they will have to shut down the air conditioning or the heating of the passenger compartment to be sure to have enough driving range and save some precious energy from the battery. There is some truth in this, but once again, it is important to consider the exact situation. Current hybrid or internal combustion engine vehicles have a very limited thermally insulated passenger compartment, even if there is some thermal insulation. Most of the insulating material used on these vehicles is for noise insulation, not for thermal insulation. The need for noise insulation is lowered on vehicles without an internal combustion engine. In the case of battery electric vehicles, a good thermal insulation concept helps to reduce the climatic dependency. An elegant technological solution is the use of thermal pre-conditioning, which consists of heating or cooling the passenger compartment and other parts in the vehicle when it is parked at the charging station, by using the electric energy coming directly from the grid in an efficient way (e.g., by powering a reversible heat pump). The thermal energy stored in the vehicle will then be available and sufficient for the first 10–20 km. Further, the thermal energy stored and produced in the different parts of the powertrain can be reused and transported where it is needed. The technological solutions exist, but they are currently implemented in very rudimentary ways. In pure battery electric vehicles, the electric, thermal and mechanical energies will have to be managed in a holistic way at the whole-vehicle level (i.e., not only on subsystem levels).

4.3 Lower Costs

Vehicles with a high level of electrification, compared with internal combustion engine vehicles, have still today to overcome high development costs compared to the volumes produced. These vehicles are mainly more expensive because of their high energy lithium-ion battery system. As of 2013, automotive lithium-ion batteries are produced in series but still not in high volumes. However, the costs of the production of battery cells and battery systems will further drop with a higher amount of automation on the production lines. But the power and control electronics have a part of the responsibility for the higher costs in electric vehicles. Currently, the costs in electronics could be further optimized through modular and more standardized parts in power electronic modules. Further, since the transmission is different in electric vehicles compared to internal combustion vehicles, these parts are costly because they are used only in a small amount of vehicles (i.e., no economies of scale exist).

In 2011 in [17–19], it was found out that plug-in hybrid electric vehicles do not compensate for their higher purchase prices, even when considered over their complete lifetime. The study was performed in the USA by considering no subsidies from the government. The study shows that plug-in hybrid electric vehicles are more expensive than pure battery electric vehicles in nearly all the comparison scenarios. This result is explained by the fact that pure battery electric vehicles are less complex to design and fabricate compared to the most complex plug-in hybrid electric vehicles having both a battery system and an internal combustion engine. An important point to mention is that the reliability was not taken into account for the investigation of the costs.

4.4 Enablers and Vision

In summary, the electrification of the current internal combustion engine vehicles adds costs. However, the achievement of a full electrification of the powertrain will further lower the final costs when compared to hybrid or pure internal combustion engine vehicles. When produced in high volumes, pure battery electric vehicles can be produced at lower costs than today's plug-in hybrid or pure internal combustion engine vehicles. The paradox is that for reducing the production costs of pure battery electric vehicles, the market demand must increase; but for the market demand to increase, the pure battery electric vehicles must be offered at lower costs. Additionally, for the market demand to increase, the customers must be convinced of the battery technology perceived by the potential customers as unreliable: For this to happen, pure battery electric vehicles must become more common and not stay just a rarity. One way to promote the battery electric vehicle could be an electric car race series, such as FIA Formula E, whereby high profile mass media coverage changes perceptions and encourages customers to believe in electric vehicle technology and trust the performance of their batteries.

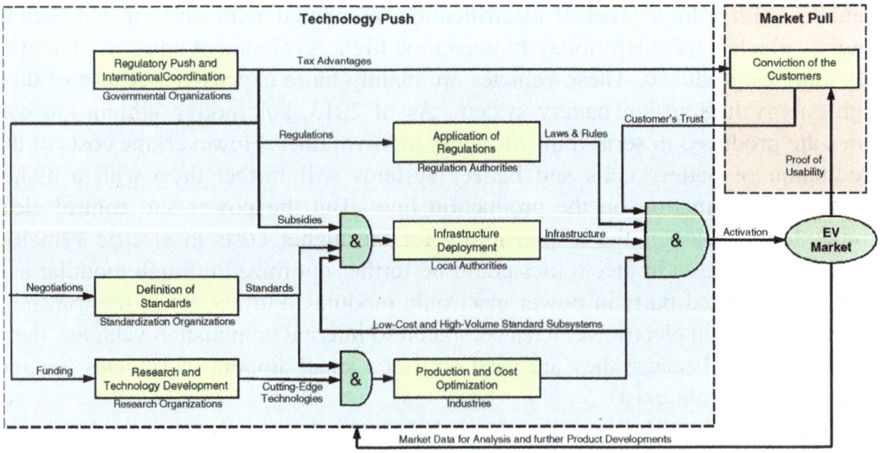

Fig. 3 Global summary of the interactions between the key enablers of the mass market of the electric vehicle. The output of the boolean logic AND-gates (i.e., "&") can only be provided when all the conditional inputs are present and satisfied

5 Conclusion

Because the internal combustion engine vehicle exists at low costs and is very widespread today, the battery electric vehicle is not successful on the mass market. The situation would probably look very different if the majority of the vehicles currently on the road were pure battery electric vehicles and a new invention called the *internal combustion engine vehicle* was presented, requiring a highly flammable liquid fuel, emitting a lot of noxious and environmentally harmful gases, requiring complex and expensive engine maintenance and containing dangerous substances, further needing new liquid fuel infrastructures with adapted road transport vehicles with the risk of accident and the consequences it implies. In 2013, the problem of the electric vehicles is no more a pure technological challenge: The main problem is the well-established internal combustion engine vehicle and its well-adapted liquid fuel infrastructure: *Idealism is not a mass market driver*. Therefore, the best way to enter the market and establish a market for the electric vehicle on a solid basis will certainly not be to push the electric vehicles to compete directly against the well-established internal combustion engine premium vehicles.

In the "Drive Green 2020: More Hope than Reality" report [17], different factors influencing the sales of electric vehicles (e.g., market trends, regulatory environment, consumer sentiment, technology development) were analyzed. It will not be easy to convince the customers to switch from internal combustion engine vehicles to hybrid or battery electric vehicles, especially because of the perceived uncertainties in these types of vehicles, like their look, their performances, their complexity and their reliability. The report states that at least one of the following conditions would be required:

- Significant increase in the price of crude oil
- Substantial breakthrough in technologies that would reduce the costs and improve customer confidence
- Coordinated government actions to push the customers to purchase electric vehicles.

By considering the information available today, these conditions seem to be unlikely to be fulfilled during the next years. However, these conditions are somehow biased. In the present work, the conducted analysis has shown that almost all the required technology breakthroughs for enabling an electric vehicle with usable good performances have already been developed in the past, but in general for other types of applications (e.g., military, space). For a use in the mass market where low costs are mandatory, these technologies will have to be optimized for low cost and high volume production. This is a common industrialization process, coming with its challenges, but currently not acting as a barrier.

A key to the success of the electric vehicle resides in the management of the electric, thermal and mechanical energies in a holistic way on vehicle level, and no longer exclusively on subsystem or component levels. The electric vehicles must definitely be designed and born electric. This will fundamentally change the vehicle topology used today in common combustion engine vehicles or hybrid electric vehicles.

The development and cost optimization of the required technological solutions must be coordinated with the development of the infrastructure, because without the suitable infrastructure, the electric vehicle market will not take off. Central roles will be played by the governments, the research organizations, the industries and the standardization organizations. A strong need for both national and international synchronization, including coordination, exists and will ultimately be mandatory.

Furthermore, a strong promotion of electric vehicles must be undertaken to change the mind of the potential customers and to establish a confident relationship between the customers and the electric technology for automotive powertrains. The current skepticism of the potential customers of electric vehicles about lithium-ion battery technologies and their related reliability must be overcome: It is currently a real obstacle for the development of the electric vehicle market. No potential customer wants to invest in an expensive battery pack without knowing how long it will last under daily usage.

There is little doubt that first of all, today, a technology push is necessary to bring the electric vehicle to the mass market. However, this technology push should not come from the research and technology development organizations, but directly from the governmental authorities. For this, new stable and realistic regulations should be defined in the long-term. These must be planed carefully with a concrete corresponding roadmap and not change with each government election, coherent with the objectives of the reduction of pollutant emissions and compatible with the available energies in the concerned countries.

Furthermore, it must be clear that the field "Mobility and Transportation" is extremely closely linked to the field "Energy". Not only at a national level, but also at an international level, coherent global policy is urgently required to address both energy and transportation in parallel ways. In almost all the developed countries, including Europe, facing the same energy and transportation challenges, such policies are not combined and not conducted in coherent parallel ways, thus ending in a *paradoxical activity*, like for example pushing electric vehicles using electricity produced by coal power plants [20]. Like the holistic electro-thermo-mechanical energy management needed in electric vehicles, a holistic energy and transportation policy is also required. The biggest successful changes in the history of the international transportation area were done after a strong coordination work between the involved countries: *Think global*. Interestingly, when considering the electric vehicle market, the nations are mainly competing against each others. A better coordination will be enabled by for example setting new international standards. This would be very wise, especially in preparing for the competition with Asian automotive manufacturers.

As summarized in Fig. 3, there is definitely not a single element or a single impulse that will be sufficient to enable the mass market for electric vehicles. If this had been the case, this would have occurred sometimes in the last 100 years.

Acknowledgments The research leading to these results has received funding from the European Union Seventh Framework Program (FP7/2007–2013) under the grant agreements No. 608770 ("eDAS"), No. 285224 ("SuperLIB"), No. 285739 ("ESTRELIA"), No. 314128 ("AVTR"), No. 260176 ("CASTOR") and from the ENIAC JU under grant agreement No. 270693-2 ("MotorBrain").

References

1. Ehsani M, Gao Y, Miller JM (2007) Hybrid electric vehicles: architecture and motor drives. Proc IEEE 95(4):719–728
2. Christensen CM, Raynor ME (2003) The innovators solution: creating and sustaining successful growth. Harvard Business Press, Boston
3. Frigant V (2011) French mega-suppliers' trajectories during the modular era: some evidences on Faurecia, Valeo and Plastic Omnium. Technical report, Groupe de Recherche en Economie Théorique et Appliquée
4. Eberle U, Müller B, von Helmolt R (2012) Fuel cell electric vehicles and hydrogen infrastructure: status 2012. Energy Environ Sci 5(10):8780–8798
5. Romm JJ (2004) The hype about hydrogen: fact and fiction in the race to save the climate. Island Press, Washington
6. Bossel U (2006) Does a hydrogen economy make sense? Proc IEEE 94(10):1826–1837
7. Thomas C (2009) Fuel cell and battery electric vehicles compared. Int J Hydrogen Energy 34 (15):6005–6020
8. Porter ME (1998) Competitive advantage: creating and sustaining superior performance. Free Press, New York
9. Brem A, Voigt K-I (2009) Integration of market pull and technology push in the corporate front end and innovation management—insights from the German software industry. Technovation 29(5):351–367
10. Gerpott TJ (2005) Strategisches technologie-und innovationsmanagement. Schäffer-Poeschel Stuttgart, Boston
11. Munro H, Noori H (1988) Measuring commitment to new manufacturing technology: integrating technological push and marketing pull concepts. IEEE Trans Eng Manage 35 (2):63–70
12. Hauschildt J, Salomo S (2011) Innovationsmanagement. Vahlen, Munich
13. Burgelman R, Sayles L (2004) Transforming invention into innovation: the conceptualization stage. Strategic Management of Technology and Innovation. McGraw-Hill, Boston, pp 682–690
14. Pfeiffer W (1997) Funktionalmarkt-Konzept zum strategischen Management prinzipieller technologischer Innovationen, vol 28. Vandenhoeck & Ruprecht, Göttingen
15. Horbach J, Rammer C, Rennings K (2012) Determinants of eco-innovations by type of environmental impact—the role of regulatory push/pull, technology push and market pull. Ecol Econ 78:112–122
16. Christensen CM (1997) The innovator's dilemma: when new technologies cause great firms to fail. Harvard Business Press, Boston
17. Power J and Associates (2010) Drive green 2020: more hope than reality? Technical report
18. Power J and Associates (2010) Future global market demand for hybrid and battery electric vehicles may be over-hyped; wild card is China. Technical report
19. Lee H, Lovellette G (2011) Will electric cars transform the US vehicle market? An analysis of the key determinants
20. Schreiber A, Zapp P, Kuckshinrichs W (2009) Environmental assessment of german electricity generation from coal-fired power plants with amine-based carbon capture. Int J Life Cycle Assess 14(6):547–559

Part II
Recharging

Identification of Market Models and Associated Billing Strategies for the Provision of EV Charging Services

Annelies Delnooz and Daan Six

Abstract The creation of an attractive market for the provision of charging services is crucial for the mass introduction of electric vehicles. Therefore there must be an adequate availability of charging infrastructure for owners of electric vehicles at their preferred locations where affordable and user-friendly charging services are offered. In this paper a part of the work performed within the ENEVATE project is described. Market models are presented that describe the structure of the market for charging services in terms of roles which can be taken up by different actors. Subsequently, this paper analyses the state of the art on billing structures for charging services. Finally, the correlation between the payment method and different factors are looked at (e.g. location of charging, parking and mobility policies and the type of end user).

Keywords Billing models · Clearing house · EV stakeholders · Market models

1 Introduction

Electric vehicles (EVs) have the potential to be a sustainable means of transportation. The commercial success, however, heavily depends on the customer. A lot of effort will be needed with regard to accessible market models, payment methods and transparent billing tariffs to make sure the EV-concept will be accepted by the end-user.

The ENEVATE project receives funding by the European Interreg Program.

A. Delnooz (✉) · D. Six
Energytechnology, VITO/Energyville, Boeretang 200, 2400 Mol, Dennenstraat 7, 3600 Genk, Belgium
e-mail: annelies.delnooz@vito.be

D. Six
e-mail: daan.six@vito.be

D. Beeton and G. Meyer (eds.), *Electric Vehicle Business Models*, Lecture Notes in Mobility, DOI 10.1007/978-3-319-12244-1_4

The ENEVATE project (European Network of Electric Vehicles and Transferring Expertise), was created specifically to facilitate the sharing of E-mobility related knowledge and experience across North West Europe (see Fig. 1). The partnership aims to [1];

- Accelerate the development of "new" supply chains
- Accelerate the development of sustainable electric vehicle charging infrastructure
- Explore the opportunities for, and implications of new E-mobility concepts
- Use research into regional pilot projects and supply chains to create an E-mobility roadmap
- Accelerate E-mobility innovation by stimulating technology partnerships and establishing transnational co-operation

The work presented in this paper is part of a toolkit designed for sharing expertise and knowledge with regard to EV infrastructure developments [2]. More specifically, this study presents an overview of potential market models applicable for the different charging locations, i.e. public and private charging places and charging at home, which can be found in Sect. 2. Section 3 looks at the various billing structures and tariffs for EV owners. Potential correlating factors are looked at in Sect. 4. Finally, Sect. 5 gives some general conclusions.

2 Market Models

This section provides an overview of the different stakeholders that take part in the EV ecosystem and outlines their expectations. In economic theory, a distinction can be made between two market models: on the one hand the neo-classical free market model and on the other hand the planned economy where government is an important actor. In the market of charging services and EV equipment operation as well different models or hybrid forms of these models co-exist. As there are currently no regulatory initiatives in this domain, different market models, pricing mechanisms and business models are being rolled out depending on the place of charging: a public domain, a private parking lot or at home.

2.1 Charging Infrastructure in the Public Domain

A significant amount of vehicle owners will have the opportunity to charge their EV at home (private garage or driveway). However, from a future perspective with better integration into the electricity grid, it is recommended to charge the battery every time the vehicle is parked. An extensive network of public charging infrastructure is in this case essential.

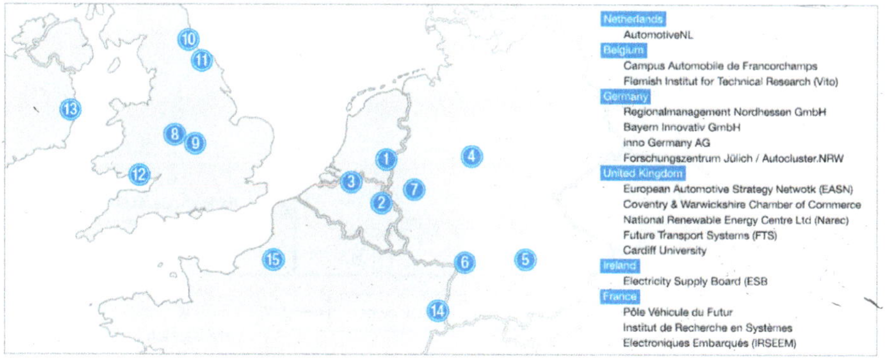

Fig. 1 Partners within the ENEVATE project

For public areas, the city or municipality is the administrator of the domain. Different market models emerge: the city or municipality can decide to offer the charging equipment as a public service to its citizens. It can either decide to outsource the operations and maintenance to a private EV equipment operator or to purchase and manage the equipment itself. In an open market model the city or municipality will not take this initiative and the market will be operated by several service providers. In some countries the distribution system operators (DSO) or the energy providers take up an active role in this new market. In Belgium the DSO assist cities and municipalities in implementing charging infrastructure. Also in Germany and the Netherlands the energy sector is taking up a central role.

In Belgium, the general policy with regard to contracting public parking operators comprises a tendering process or the attribution of a concession to a private operator. The local governments, in both procedures, can define the parking fees since the tariff responsibility remains with the city or municipality.

For the installation of EV charging infrastructure in the public domain, a comparable tendering process or concession agreement can be used.

The EV service provider, the legal entity that the customer has a contract with for charging services, has to obtain information about the charging transactions from different public charging networks, operated by several EV equipment operators. In order to enable the exchange of information, the EV service providers will have to sign bilateral agreements with each EV equipment operator. To facilitate this process, an intermediary clearing house which acts as the sole clearing counterparty can assist (see Fig. 2). An example of such a clearing house is e-clearing.net in the Netherlands [3].

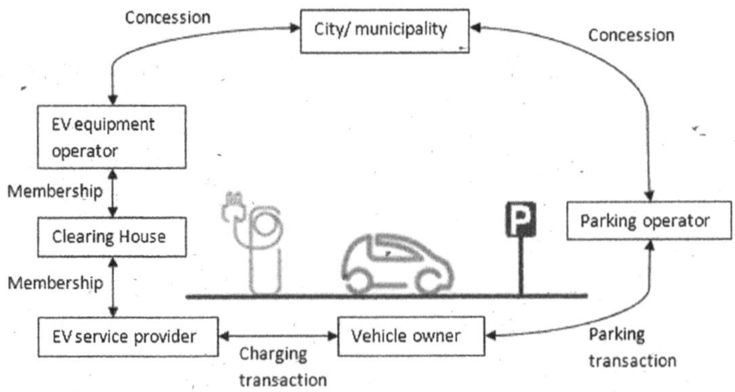

Fig. 2 Market proposition for public charging infrastructure

2.2 *Private Charging Infrastructure*

Besides public charging infrastructure, vehicle owners can also turn to private charging spots when located away from home. An example of such a private charging infrastructure can be found in the Flemish Living Lab Electric Vehicles [4], where the supermarket chain Lidl, offers charging facilities for bikes and cars at two of its parking lots. For the operation and maintenance of the charging equipment, the domain owner, Lidl, relies on an EV equipment operator under a service agreement.

The access of the different EV service providers on privately operated charging infrastructure is also accomplished by a Clearing House (see Fig. 3). As the number of EV service providers and EV equipment operators increases, the corresponding costs of managing the bilateral contracts will augment as well.

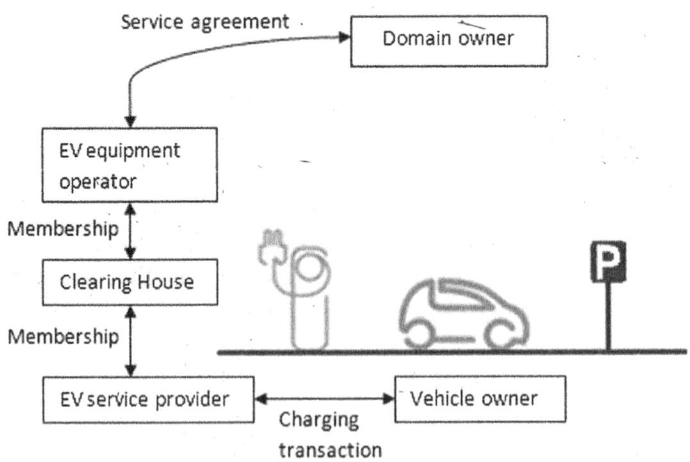

Fig. 3 Market proposition for private charging infrastructure

2.3 Home Charging Infrastructure

When looking at the home location, the EV owner can opt to charge his vehicle via a conventional electricity socket or install a dedicated charging equipment, a so called wallbox.

The electricity delivered to the home power connection can be supplied via a contract agreement with an EV service provider or via a contract with his standard electricity supplier.

If charging via the standard electricity provider is preferred, the EV owner can opt to enter into a supplementary contract with his electricity supplier for the electricity provision for his electric vehicle. Afterwards the EV owner receives an invoice for the extracted energy. The invoice can either be a joint invoice which bundles his domestic electricity consumption and his electricity consumption for his electric vehicle or a separate invoice for both types of electricity consumption. However, the electricity consumption of the EV can only be individually metered if a dedicated wallbox is installed which is equipped with a separate electricity meter. The EV charging equipment is obtained by the vehicle owner via either a leasing agreement, or it is bought or hired from an EV equipment operator (see Fig. 4).

The EV owner can also opt to enter into a service contract with an EV service provider. In this case, a dedicated wallbox needs to be installed by an EV equipment operator with a separate electricity meter. The installed wallbox must receive an individual EAN-number since it is currently not legally permitted, in Belgium, to have two energy suppliers assigned to one EAN-connection. It can be that the energy supplier of the EV owner, which supplies the domestic electricity, is not the same as the contracted energy supplier of the EV equipment operator. This emphasizes the importance of a separate EAN-number since the EV equipment operator must have the ability to choose the energy supplier as administrator of the EV equipment.

The Single Point Of Contact (SPOC) of the vehicle owner is the EV service provider, who offers charging services to the end-customer. This implies that the vehicle owner does not know which energy supplier is actually delivering the electricity to his home charging box (see Fig. 5).

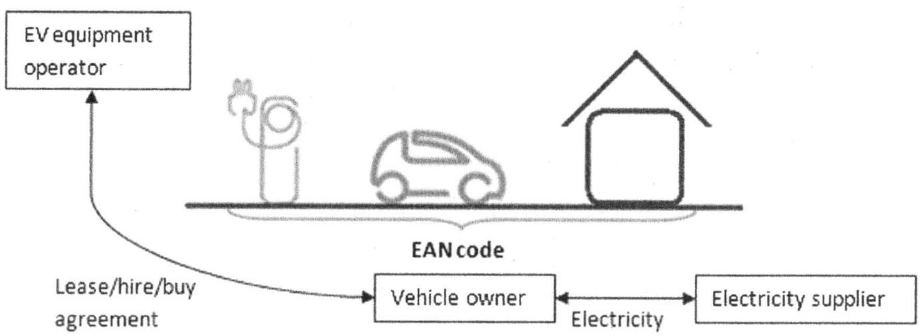

Fig. 4 Market proposition for charging at home via standard electricity supplier

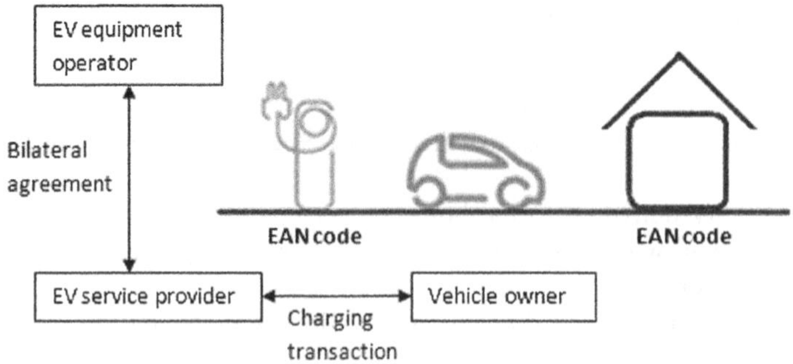

Fig. 5 Market proposition for charging at home via an EV service provider

3 Billing Structures

When looking at the key factors influencing the consumer to potentially purchase an electric vehicle, cost related components, including the cost of usage, top the list [5]. For the mass introduction of electric mobility, billing strategies must thus be carefully drafted. The starting point for the chosen billing method must be that the payment system is accessible, safe for the user and operator, resistant to fraud and little susceptible to interference. The transaction cost of the various payment methods must also be taken into account.

In the following section a suggestion is made of the various payment methods, followed by a brief introduction to different billing rates.

3.1 Payment Method

The simplest billing strategy is unmetered, free charging where there is no money transfer from the EV-owner for the purpose of the charging services received. In practice, different stakeholders (e.g. municipalities, cities, commercial businesses, pilot projects) are often offering electricity free of charge to vehicle owners. Marketing purposes lie at the basis of this free service of commercial businesses like Lidl, McDonalds, Holiday Inn and Waasland Shopping Centre. Also in many cities and municipalities, vehicle owners can benefit from free electricity for charging during the introduction phase of e-mobility. However, it is not expected that local governments and other stakeholders can keep offering free charging when the number of electric vehicles increases. The city of Amsterdam, for example, has already abandoned the principle of free charging from April 2012.

Concerning the remaining payment methods, a distinction can be made between two main types. Firstly electricity used for charging electric vehicles can be paid for before the actual charging process takes place. This category of payment methods is referred to as 'prepaid' since it entails a financial transaction pre-charging.

The second category refers to the payment after the charging process has taken place. This payment method is better known as 'postpaid' and relates to payments via cash, card, billing and by mobile. Post paid methods for billing are more complex, in the sense that they require more data communication and hence entail more software costs.

Besides the two main payment categories, the vehicle owner can also be billed for the charging services by all kinds of combinations of prepaid and postpaid methods.

3.1.1 Prepaid Methods

If the customer makes the choice to pay before the actual charging process, he will be confronted with one of the following payment methods.

(a) Subscription
The first prepaid payment method exists of unmetered energy use. The owner of the electric vehicle pays a fixed amount in order to be given access to certain charging points for a certain period of time, mostly 6 or 12 months [6]. This is a very simple payment method since the identification of the subscribed user is the only requirement. However, this method also entails some major issues. In particular, some users will be paying too much for their electricity used for charging while others will be paying far too little. It can be stated that this payment method is not beneficial for the energy efficiency since all benefits of energy efficient driving are thrown away by subscription.

(b) Pay as you go (PAYG)
In this case, the owner of the electric vehicle pays in advance to obtain a level of credit. After charging, this credit is debited and the remaining balance is determined. This payment method follows a similar approach as the prepaid cards for mobile phones. There are two potential ways in which the relevant credit can be treated [6];

- At the start of the charging, the charging station communicates with the EV service provider to verify the owners' identity and the remaining balance of the payment card. After the charging, the data concerning the extracted electricity is sent back to the EV service provider, who recalculates the remaining credit.
- The data concerning the owners' identification and the level of credit is stored on the device for identification (e.g. RFID card, mobile phone for NFC). After charging the authentication device is updated.

Despite the fact that it is a more complex and costly charging system, this method of payment is already being used by different operators of charging infrastructure (e.g. Elektromotive, POD Point, Ville de Paris). There are also some means of payment on the market who fit the 'Pay as you go'-principle e.g. OV-chipcard, Chipknip and PingPing [7]. The PingPing platform of

Mobile For relies on a prepaid mobile wallet, linked to a back-office system. The virtual wallet can be recharged via the bank account (internet banking), maestro or even by a third party (e.g. employer) [8].

(c) Cash

Cash payments can also be seen as a prepaid method. Within this billing strategy the EV owner is expected to insert cash into the appropriate device on which the meter of the relevant charging infrastructure starts running and the respective amount of electricity, in accordance to the cash payment, is transferred to the EV.

3.1.2 Postpaid Methods

Next to prepaid methods, electricity used for charging electric vehicles can be paid for after the actual charging process has taken place. The available billing concepts within this category are listed below.

(a) Cash

The owner of the electric vehicle pays for the electricity by using cash. This payment method shows some similarities with the current payment method at petrol stations. In Belgium, Total is experimenting with inter alia, cash payments for charging within its PlugToDrive network [9]. However, while the value of the electricity used for charging is rather small, the costs for collecting, storing and employing somebody to retrieve the cash are relatively high. It will thus be cost ineffectual to give customers the opportunity to pay using cash.

(b) Card

In this case, the owner of the electric vehicle pays for the electricity by using his credit card. Also this method shows some similarities to the current payment method at petrol stations.

(c) Pay by mobile

Within this model the owner of the electric vehicle sends an SMS to a dedicated number, indicated by the EV equipment operator. The telecom operator of the EV equipment operator then charges the mobile phone operator of the EV owner with the amount applicable.

However it is very convenient to pay by mobile phone, this system is not a plug-and-play system. This payment method entails additional administration costs as there needs to be extensive communication between both the EV service provider and the mobile phone companies of both the EV equipment operator and the EV owner.

(d) Domestic electricity bill

Since most customers already have a domestic electricity account, it may be convenient for owners of EVs to have their electricity usage for the charging of the EV added to the domestic electricity bill. This would require that for public charging points both the used electricity and the user authentication must be sent to the electricity supplier. In this case there will be a great need for a Clearing House since the standard electricity supplier of the vehicle owner is not necessarily the dedicated electricity supplier of the public charging point.

For home charging stations, the current domestic electricity meter can be used to measure the electricity usage for charging the electric vehicle.

(e) Separate electric vehicle electricity bill

This payment method exhibits some similarities to the above payment model but in this case the owner of the electric vehicle receives a separate invoice for the electricity used for charging the EV. This entails that the customer needs to install a separate meters at home for the domestic electricity use and for EV charging. Within this payment method the owner will receive a separate electricity account for the charging of the electric vehicle.

If the vehicle owner (also) charges at public charging points, the electricity used can be included on an integrated EV electricity bill if a Clearing House takes care of the necessary clearing transactions.

3.1.3 Combined Billing Structures

Besides the two standard payment methods, prepaid and postpaid billing, the vehicle owner can also be billed for the purchase of charging services via different combined billing structures.

For instance; the owner of the EV can have a billing contract with an EV service provider where he pays a fixed amount each month and retrieves a certain contracted energy use in return. If the user extracts more electricity than the set amount, he will be charged on top of the contracted amount.

Another example of a combination between prepaid and postpaid pricing can be found in the billing schemes of Blue Corner, a Belgian EV service provider [10]. One of the formulas offered by Blue Corner consists of a yearly subscription fee and additionally vehicle owners need to pay a use fee after each charging session.

3.2 Billing Rate

In all of the above payment methods, it is possible to either charge the user per kWh (real energy flow based), per unit of time connected or per charging session.

- Time-based: the owner of the EV pays a certain amount to be connected to the charging point for a certain period of time. During the charging session the real energy flow can vary depending on inter alia, the state of the battery.

- Real energy flow based (kWh): most commonly, charging rates are expressed per kWh. Then the owner of the electric vehicle is charged per kWh of energy he extracts from the electricity grid. Many operators of charging infrastructure are experimenting with Time-of-Use rates where electricity is offered at different prices based on the time of the day when the electricity is actually used.
- The vehicle owner can also be billed per charging session. In this case neither the time connected, nor the real energy flow is monitored. For this tariff structure, the critical parameter is the number of times connected to a charging infrastructure.

4 Correlated Factors

4.1 Location of Charging

Not all payment methods previously mentioned are suited for all types of charging infrastructure, domestic (home charging infrastructure) and non-domestic (public and private charging infrastructure).

Generally, it can be stated that most non-domestic charging points will probably be equipped with facilities for postpaid payment as it is a more flexible billing type. Furthermore, the integration of parking fees in the overall charging tariff requires postpaid payment methods.

For domestic charging infrastructure, the electricity will probably be billed by a contract, irrespective of whether postpaid or prepaid (e.g. subscription, domestic/ separate electricity bill). The payment method most likely to be used in a domestic location is to charge consumers for the refueling of electric vehicles by their domestic electricity bill. This method is the easiest to use or implement for both the consumer and the electricity supplier since most consumers already have a metering and billing system in place for their domestic electricity use and this method expects nothing new from the consumer.

For both locations the payment by cash is expected to be the least likely to be implemented since it is inconvenient for the consumer and difficult to implement by both the EV equipment operator and the EV service provider. While the value of the electricity used for charging is low, the costs for collecting, storing and employing somebody to retrieve the cash are relatively high.

4.2 Parking and Mobility Policies

The parking and mobility policies of the local governments are closely linked to the success of e-mobility in the relevant municipality or city as they are in the position to facilitate or thwart the use of electric vehicles.

In the context of a parking policy, the local government grants public parking areas under a concession agreement. Since it always concerns a concession of public service, meaning the local government can set the parking tariff, the city or municipality has a direct impact not only on the availability, but also on the affordability of public parking spots. Consequently there is an impact of the parking policy on the availability and affordability of charging an EV in the public domain.

Furthermore it should be noted that if the parking fee is integrated into the general tariff for charging, one must strive towards a time-based billing structure. The reason being that the value of parking is greater than the value of the electricity for the EV owner, especially in urban areas. If the EV owner is billed based on the real energy flow or per charging session, he can have the incentive to park longer than necessary for the charging process.

Within the framework of a local mobility plan, cities and municipalities often work out a congestion management plan for traffic in city centers to manage or even ban vehicles. This could result in EVs are not being allowed into inner city areas and charging infrastructure being installed in less convenient locations.

4.3 Type of End User

A large proportion of the total vehicle fleet consists of company cars that are subject to a leasing agreement of the employer. These specific types of vehicles require additional attention when considering EVs that need to be electrically charged. In particular, the employer wants to be assured that he is only paying for the charging services of the leased vehicle. This entails that the electricity consumption of the leased EV has to be metered separately. In non-domestic locations, the public and private charging infrastructure is already equipped with a metering system and thus charging at these locations is less of an issue.

A more critical location, when it concerns the separate metering of the leased EV, is the domestic location. At home, the EV owner has the ability to charge his company vehicle via a conventional electricity socket. In this case the electricity consumption of the vehicle cannot be individually metered and is incorporated in the general domestic electricity consumption. The installation of dedicated charging equipment, a so-called wallbox, is essential in this context.

In addition to the choice of the means by which the vehicle is charged (standard electricity socket or dedicated charging equipment), the delivered electricity to the home power connection can be supplied via a contract agreement with an EV service provider or via a contract with his standard electricity provider.

In the case of a leased company vehicle, it is preferred to rely on an EV service provider. Since, in the case that the delivered electricity is provided by the standard electricity supplier, it sets high requirements for clearing. In particular, the electricity consumption of the company vehicle must be subtracted from the domestic electricity bill, or even from the separate EV bill in the case more EVs are charged at home. The information on the extracted electricity and the associated charges must be shared between the standard electricity supplier of the vehicle owner and the electricity supplier or service provider of the employer since the latter is the paying party. If the charging service is delivered by an EV service provider, the organizational structure and data communication are less complex. However, in this case the installed wallbox must receive an individual EAN-number.

5 Conclusions

This paper provides an overview of potential market models for the provision of EV charging services looking both at free-market models and planned economy designs. As there are presently no regulatory initiatives in this domain, different market models, pricing mechanisms and business models are being rolled out depending on the place of charging: a public domain, a private parking lot or at home.

Subsequently, this paper analyses the state of the art on billing structures for charging services. A description is provided of the potential payment methods focussing on prepaid and postpaid methods as well as various combined billing structures. This analysis is followed by a brief introduction to different billing rates.

Finally, the correlation between the payment method and different factors are looked at (e.g. location of charging, parking and mobility policies and the type of end user).

References

1. ENEVATE project. http://www.ENEVATE.eu
2. Lumsden M, Electric vehicle charging infrastructure tool kit. Future Transport Systems Ltd, ENEVATE. http://www.enevate.eu/fb355ab2-6442-c102-fe5f-a586908aef00?Edition=en
3. E-clearing.net. http://www.e-clearing.eu/
4. Flemish living lab electric vehicles. http://www.livinglab-ev.be/content/introduction
5. Nieuwenhuis P (2012) Which EV experiments are likely to succeed, which are not? Milestone conference accelerating e-mobility, Cardiff University, ENEVATE, 15 May 2012
6. Bending S, Channon S, Ferdowsi M, Nadoli F, Bower E (2010) Deliverable 1.1: specifications for EV-grid interfacing, communication and smart metering technologies, including traffic patterns and human behavior description, MERGE, August 2010
7. Accenture (2010) Studie marktmodel laadinfrastructuur ten behoeve van elektrisch vervoer, April 2010
8. Jacobs S Toelichting nieuwe visie en focus walstroombetaalsysteem, Mobile-For, Oktober 2012
9. Total Belgium, Elektrische voertuigen, Total Belgium pioniert. http://www.total.be/nl/stations/brandstoffen/plug-to-drive.html
10. Blue Corner, Abonnementen: formules. http://www.bluecorner.be/abonnementen_overzicht.html

Business Case for EV Charging on the Motorway Network in Denmark

Victor Hug

Abstract This paper presents a business case for the development and operation of EV (Electric Vehicle) fast charging stations on the Danish motorway network. The business case is based upon an analysis of the EV models available on the Danish market, existing high power recharging solutions and a projection of market uptake of EVs in Denmark. The business case investigates two scenarios: A low scenario where 5 % of an EV's charging demand is covered by fast charging stations and a high scenario where fast charging stations cover 10 % of an EV's charging requirements. The two scenarios are examined for both a 5-year and a 10-year period. The business case shows that fast charging stations throughout a period of 10 years may result in a significant return of investments whereas investments for a period of only 5 years is too risky.

Keywords Electric vehicle charging · Charging station · Business case · Motorway network · Denmark

1 Background

In January 2013, the European Commission proposed a directive on deployment of an alternative fuels infrastructure. The proposed directive sets up national targets for public accessible EV charging points and proposes a European standard for EV fast charging [1].

In response to the proposed directive, an inter-ministerial working group including the Danish Energy Agency, the Danish Transport Authorities and the Danish Road Directorate analyzed the financial, legal and technical aspects of establishing an infrastructure for alternative fuels on the Danish motorway network.

V. Hug (✉)
Victor Hug Consult, Granlien 8, 2400 København NV, Denmark
e-mail: victor.c.hug@gmail.com

D. Beeton and G. Meyer (eds.), *Electric Vehicle Business Models*,
Lecture Notes in Mobility, DOI 10.1007/978-3-319-12244-1_5

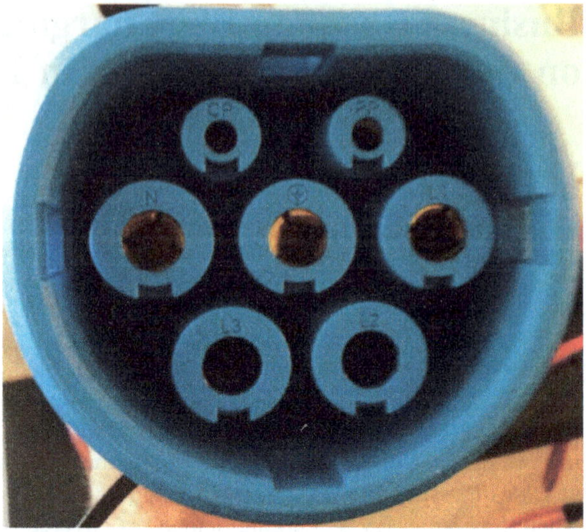

Fig. 1 Type 2 plug (vehicle connector)

Based on the working group's recommendations, the Danish Transport Ministry and the Danish Ministry of Climate, Energy and Buildings have decided to carry out a public tender on publicly available charging stations at service stations at the Danish motorway network [2]. The tender process is planned for the spring of 2014.[1]

As input to the working group, Victor Hug Consult conducted a study for the Danish Energy Agency on existing technologies applicable for fast charging of EVs and developed a business case for potential revenue from such recharging stations [3]. This paper is based on this study.

2 EV Charging Solutions Applied in Denmark

The EV battery is always charged with Direct Current (DC). Alternating Current (AC) is either converted in the charging station (DC charging) or in a vehicle on-board converter (AC charging).

One factor that is crucial for the success of EVs is a uniform system for EV charging across Europe. Denmark, along with most European countries, decided to use the Type 2 standard for AC charging (Figs. 1 and 2). The Type 2 standard supports three-phase charging. The Danish power supply is based on a three-phase AC power system where the consumer has access to either one phase with 230 V or three phases with 400 V. EV home charging facilities is normally restricted to 11 kW AC (three phases, 400 V and 16 A). Public AC charging stations presently

[1] The Danish Road Directorate is organizing the tender process.

Fig. 2 Type 2 socket (vehicle inlet)

support up to 22 kW AC (three phases, 400 V and 32 A) and can potentially support up to 43 kW AC (three phases, 400 V and 63 A).

DC charging in Denmark is today based on the CHAdeMO and the Combined Charging System (CCS) standards. The CHAdeMO standard is adopted by the Japanese industry and supports up to 50 kW charging (Figs. 3 and 4).

U.S. and German car manufacturers have adopted another standard for DC charging—SAE International's Combined Charging System (CCS) which combines AC and DC charging in one socket. In Europe this socket is called Type 2 Combo (or Combo 2). The Type 2 Combo socket is compatible with the Type 2 plug (Figs. 5 and 6). A vehicle equipped with a Type 2 Combo socket may be charged with DC using the Type 2 Combo plug or with AC using the Type 2 plug. The CCS protocol presently supports up to 100 kW charging. A number of EV models use CCS for DC charging. These vehicles can, however, only charge with up to 40–50 kW.

In the proposed directive on deployment of an alternative fuels infrastructure the European Commission suggests the use of Type 2 for AC charging and Type 2 Combo for DC charging.

There are no common definitions of EV charging velocity. The European Commission defines slow charging as charging with a power of up to 22 kW, whereas charging with more than 22 kW is referred to as fast charging [1]. The Clean Energy Ministerial EV Initiative (CEM EVI) defines slow charging as AC charging where a full charging takes 4–12 h, and fast charge as DC charging where a full charge takes between 0.5 and 2 h [4].

This paper investigates the use of high power charging. High power charging is defined as charging with 20 kW or higher power.

Fig. 3 Yazaki (CHAdeMO) plug (vehicle connector)

Fig. 4 Yazaki (CHAdeMO) socket (vehicle inlet)

3 The Danish EV Recharging Market

Two e-Mobility Providers (EMP), CLEVER (formerly known as ChoosEV) and E. ON, dominate the Danish market for EV charging from publicly available charging stations. In September 2013, the German utility provider E.ON took over all charging posts formerly owned by Better Place Denmark.

Fig. 5 Type 2 Combo plug (vehicle connector)

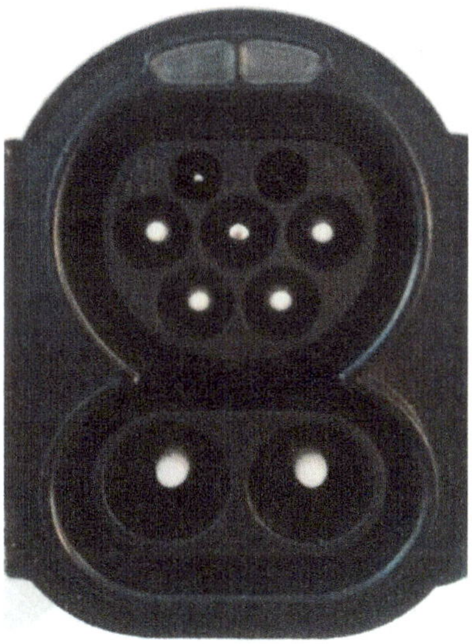

Fig. 6 Type 2 Combo socket (vehicle inlet)

During 2012, Better Place and CLEVER had launched two country-wide networks for fast recharging of EVs. Better Place's EV recharging network was based on Battery Swap Stations (BSS) whereas CLEVER's charging network use high power AC and DC charging stations.

On the 26th May 2013, Better Place Denmark filed for bankruptcy. Subsequently, all 18 BSS were closed. It is not settled what will happen with the BSS,[2] but as Better Place only succeeded to sign an agreement with one EV manufacturer (Renault) on production of one battery-switch ready model (Fluence Z.E.), Better Place's Battery Swap business model does not seem to be financially sustainable.

Besides high power DC charging stations, CLEVER's EV recharging network consists of 3.7 and 22 kW AC charging stations. E.ONs network consists of 11 kW charging posts.

By November 2013, the total number of publicly available high power charging stations in Denmark amounted to 49 publicly available 22 kW Type 2 AC charging posts (with a total of 98 socket outlets), 51 CHAdeMO 50 kW DC charging stations and 6 combined 20 kW CHAdeMO DC and 22 kW Type 2 AC charging stations [5].

Today only one charging station is located at service stations on the Danish motorway network. The charging station (50 kW CHAdeMO) belongs to CLEVER's recharging station network.

CLEVER opened its first CCS charging station in December 2013 and plan to deploy 50 CCS recharging station during 2014 and another 150 22 kW AC (Type 2) charging stations each equipped with 2 socket outlets [6].

4 Charging Specifications of EVs on the Danish Market 2013–2014

All new EV models that are introduced to the Danish market in 2013–2014 can be charged with high power AC or DC, see Table 1. A few of the existing models on the market can only be charged with up to respectively 3.7 or 6.1 kW AC (Renault Kangoo Z.E., Renault Fluence Z.E. and Mercedes Vito E-Cell).

EVs on the Danish market in 2013–2014 are produced by German, French, American and Japanese car manufactures. EVs from Nissan, Mitsubishi and PSA Peugeot Citroën are compatible with the CHAdeMO standard, EVs from BMW and VW are using CCS and EVs from Daimler and Renault can be charged with up to 22 and 43 kW AC respectively. Tesla are using the Type 2 socket inlet in the Tesla S model for both AC and DC charging. The car can be charged with up to 22 kW AC or 135 kW DC.

All EVs on the Danish market are capable of AC charging. Some EVs are equipped with a Type 1 socket inlet (the standard used by American, Japanese and some French car manufactures) and some are equipped with the Type 2 socket inlet. All EVs are delivered with an on-board charging cable. The cable is equipped with a vehicle connector that is compatible with the car's socket inlet (Type 1 or Type 2). The other end of the cable is equipped with a Type 2 plug that is compatible with the publicly available charging infrastructure.

[2] Some BSS will be torn down.

Table 1 Vehicle socket inlet and max charging power of EVs (passenger cars and light duty vehicles) on the Danish market in 2013–2014

Model	AC type 1 (kW)	AC type 2 (kW)	DC type 2 (kW)	CHAdeMO (kW)	Type 2 Combo (kW)
BMW i3		3.7[b]			50
Volkswagen e-Golf		3.7			50
Volkswagen e-Up		3.7			40
Citroën C-Zero	3.7			50	
Citroën Berlingo Electric	3.7			50	
Mitsubishi iMiEV 2013	3.7			50	
Nissan e-NV200	3.7			50	
Nissan Euro Leaf 2013	3.7[a]			50	
Peugeot iOn	3.7			50	
Peugeot Partner Electric	3.7			50	
Renault Zoe Z.E.		43			
SMART ED3		22			
Tesla S		22	135		
Mercedes Vito E-Cell		6.1			
Renault Kangoo Z.E.	3.7				

Source Victor Hug Consult [7]
Notes
[a] Is prepared for 6.6 kW AC charging (230 V and 32 A on one phase). Nissan does not offer this solution in Denmark until it is settled if this solutions is legal and technical feasible in Denmark
[b] Is prepared for 7.4 kW AC charging (230 V and 32 A on one phase). BMW does not offer this solution in Denmark until it is settled if this solutions is legal and technical feasible in Denmark

For some EV models, high power DC solutions are considered as extra equipment. It is, however, expected that almost all EV buyers will prefer to purchase a high power DC solution if possible. The largest share of EVs in Denmark is today bought by fleet owners—particularly municipalities. Even if the fleet owner do not plan to use DC charging of the vehicle in the daily operation, it is properly preferable to purchase the car with the DC charging option as the DC charging option may increase the resale value of the car significantly.[3]

5 Duration of Charging Sessions

The EV battery's capacity is specified as % State of Charge (SOC). The charging of the EV's battery is controlled by the vehicle's Battery Management System (BMS). The duration of a charging session depends on the available power (kW) and the EV's capability to absorb the power load.

[3] The price of the VW e-Up is DKK 186,000 (USD 33,818). The CCS DC charging option costs additionally DKK 4,000 (USD 727) [8].

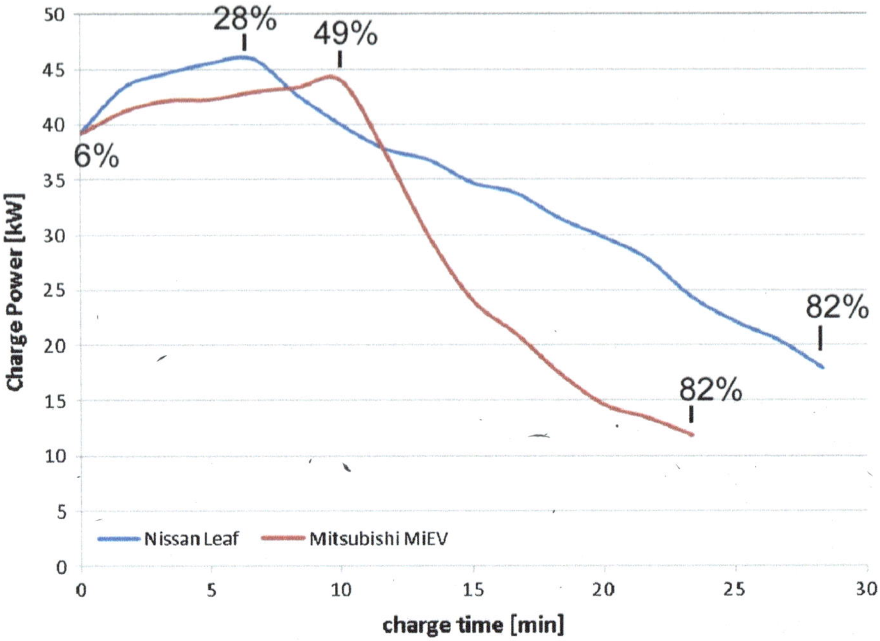

Fig. 7 Charge sessions of Nissan LEAF and Mitsubishi iMiEV from 6 to 82 % SOC. SOC at peak power is based on own calculation. *Source* ABB [10]. Courtesy of ABB

AC charging allows the EV to be charged with peak power load to above 80 % of the battery's SOC.[4] DC charging of most EV models on the other hand is only conducted by peak power for a shorter period of time.

Figure 7 illustrates the charging session for DC charging of a Nissan LEAF and a Mitsubishi iMiEV. The starting capacity of the battery is 6 % SOC and the charging session stops at 82 % SOC. Peak power for Nissan LEAF is reached after 6 min at 28 % SOC whereas peak power for Mitsubishi iMiEV occurs after 10 min at 49 % SOC. DC charging time is dependent on multiple factors including temperature. It is evident from the charging session example that the highest power of the DC charging of these two EV models takes place in the first part of the charging session and for a relatively short time period.

The battery size of a Nissan LEAF is 24 kWh. It takes 30 min to charge the vehicle from 0 to 80 % SOC with DC (peak power of 45–50 kW) [11]. Correspondingly, it takes 1 h to charge a Renault Zoe with a 22 kWh battery from 0–80 % SOC with AC (peak power of 22 kW) [12].

[4] A Danish test of the Renault Zoe shows that the vehicle can be charged with 22 kW power load until 92 % of the battery's SOC [9].

5 min of 50 kW DC charging provides an EV with power for approximately 18 km driving[5] whereas 5 min 22 kW DC charging issues an EV with enough power for around 8 km.[6]

6 Distribution of EV Charging According to Location of Charging Station

The Danish EV test program Test an EV (managed by CLEVER) encompasses 198 EVs. The EVs are loaned to families for a period of 3 months before the vehicles are passed on to new families. By the end of July 2013, 1518 families had driven 4.4 million km in the EVs and charged the vehicles more than 67.000 times. The program primarily includes the three EV models Mitsubishi iMiEV, Peugeot iOn and Citroën C-Zero which are almost identical cars.

CLEVER has analyzed the users' charging behavior. Based on more than 50,000 charging sessions, CLEVER concludes that:

- 70 % of all charging takes place at home (AC charging)
- 22 % of all charging is non-domestic AC charging (e.g. at publicly available AC recharging stations or work place charging)
- 8 % takes place at publicly available DC recharging stations.

The families have not been charged for using CLEVER's publicly available AC or DC stations. The price for using CLEVER's publicly available charging infrastructure is DKK 3.5–5.5 per kWh (USD 0.64–1) (including VAT) [15]. Private consumers typically pay DKK 2–2.5 per kWh (USD 0.36–0.45) (including VAT) for residential charging. If the test families had been charged for using CLEVER's publicly available charging, the share of home charging would probably have been higher.

ABB is today the largest provider of DC charging solutions on the Danish market. ABB expects that 5–10 % of all EV charging will take place at publicly available DC charging stations [16].

For the purpose of this article, it is expected that the power consumption for EV charging will be shared in the following way:

- 80–90 % of charging takes place at home or at work places (AC charging)
- 5–10 % takes place at publicly available AC charging stations
- 5–10 % takes place at publicly available DC charging stations.

---

5 The peak power of an ABB Terra 51 DC CHAdeMO charging station is 50 kW. The calculation is based upon a charging power of 45 kW and an average energy consumption of 200 Wh/km. Danish tests of Mitsubishi iMiEV and Renault Fluenze Z.E. show an average energy consumption of 201 Wh/km [13, 14].

6 Based upon an efficiency loss of 12 % and an average consumption of 200 Wh/km. Danish tests of Mitsubishi iMiEV shows a charging efficiency loss of 12 % [14].

Table 2 Projection of EVs and EV energy consumption 2013–2025

Year	Number of EVs	Consumption (GWh)					
		Total	Public DC		Public AC		
			Low	High	Low	High	
2013	1,391	4	0.2	0.4	0.2	0.4	
2014	3,076	9	0.5	0.9	0.5	0.9	
2015	6,047	18	0.9	1.8	0.9	1.8	
2016	10,602	32	1.6	3.2	1.6	3.2	
2017	16,842	51	2.6	5.1	2.6	5.1	
2018	24,864	75	3.8	7.5	3.8	7.5	
2019	34,768	104	5.2	10.4	5.2	10.4	
2020	46,653	140	7.0	14.0	7.0	14.0	
2021	59,529	179	9.0	17.9	9.0	17.9	
2022	73,395	220	11.0	22.0	11.0	22.0	
2023	88,249	265	13.3	26.5	13.3	26.5	
2024	104,094	312	15.6	31.2	15.6	31.2	
2025	120,927	363	18.2	36.3	18.2	36.3	

Source EN.dk [18]

For charging from publicly available recharging stations two scenario are set up: a high scenario where 10 % of all EV charging derives from high power AC and DC charging stations and a low scenario where AC and DC charging from publicly available high power charging stations constitute 5 % of the total energy consumption from EVs.

7 Market Uptake of EVs in Denmark

The Danish Energy Association (a commercial and professional organization for Danish energy companies), Energinet.dk (Danish national transmission system operator for electricity and natural gas) and Dong Energy (the largest Danish energy utility company and distribution system operator) have developed a model for market uptake of EVs in Denmark.

The model includes 3 scenarios: a high, a moderate and a low scenario, where the moderate scenario is considered most likely[7] [17]. Table 2 shows the projection of the increase in uptake of EVs until 2025 based on the moderate scenario. The EV projection model also includes energy consumption from Danish EVs. The calculation is based upon an average energy consumption per EV of 3,000 kWh/year.

Calculation of energy consumption from EVs in this article is based on the model's moderate scenario. The model is presently the most reliable projection of EVs in

[7] The model is based upon a range of parameters including total cost of ownership, available charging infrastructure and user preferences.

Denmark (some stakeholders do, however, consider the model's scenarios too optimistic).

Table 2 also shows the corresponding energy consumption from publicly available high power AC and DC charging station under assumption of the above outlined high (10 %) and low (5 %) scenarios.

8 EV Charging Station Located on the Motorway Network

It is assumed that the EV driver prefers to charge the EV at home if possible. Domestic charging is cheaper and more convenient than charging at publicly available charging stations. The EV driver is therefore expected to make an effort to reduce the use of public charging to a minimum. Hence, the EV driver will only charge the amount of power needed to get home or to arrive at another cheap and convenient charging facility (where the charging does not entail undesired waiting time). As outlined above, DC charging is especially suited to 5–10 min charging sessions, where the EV driver can quickly charge the energy needed to continue the trip.[8]

EV charging at the motorway network will also include charging for long distance trips. The EV driver will typically charge the battery up to 80–100 % SOC at one or more charging stations in order to be able to arrive at the destination.

Motorway charging infrastructure will be situated at manned service areas (equipped with gasoline stations and/or restaurants). In order to be attractive for the consumer to use, the available charging solutions should support the highest power load that the EV can adopt.

ABB and RWE (in Denmark retailed by CleanCharge) are two of the leading European manufactures of high power EV charging solutions. This analysis of EV charging solutions is based on ABB and RWE's product portfolios and EV charging stations manufactured by a small Danish charging station retailer, EVergreen.[9] A selection of available charging stations on the Danish market is shown in Table 3.

Table 4 shows the total costs associated with establishing EV high power charging stations at motorway service areas. Cost data is based on establishing a series of high power charging stations at 20 different sites. It is assumed that the distance from the charging station to the connection point of the electric grid is 25 m—hereof 10 m through paved areas and 15 m through non-paved areas.

---

[8] This does not apply for Tesla drivers, as Tesla has developed a business model where the owner of a Tesla S can charge for free at the company's network of superchargers.

[9] Other charging station providers on the Danish market refused to contribute to the study.

Table 3 Selected high power charging stations available on the Danish market

Manufacturer	Model	Type and units of plugs/socket outlets					
		Units	AC type 2 (kW)	Units	CHAdeMO (Yazaki) (kW)	Units	Type 2 Combo
RWE	eStationCITY SMART[a]	2	22	–	–	–	–
Evergreen	Terra[a]	2	22	–	–	–	–
Evergreen	Magnum[e]	4	11	–	–	–	–
ABB	Terra SC[b]	–	–	1	20	–	–
ABB	Terra SC Duo Charge[b]	1	22	1	20	–	–
ABB	Terra 51[b]	–	–	1	50	–	–
ABB	Terra 100.2	–	–	2	50	–	–
ABB	Terra 52[c]	1	22	1	50	–	–
RWE	eStation Combi[c]	1	22	1	50	–	–
ABB	Terra 53 C[b]	–	–	–	–	1	50
ABB	Terra 53 CJ[b]	–	–	1	50	1	50
ABB	Terra 53 CJG[c]	1	22	1	50	1	50

Sources Refs. [19–21]

Notes

[a] May simultaneously charge two EVs with peak power load

[b] May charge one EV at a time

[c] May charge one EV with DC and simultaneously charge another EV with AC with excess available power (load sharing)

[d] Equipped with 4 socket outlets. May charge with a total of 63 A. Different combinations are possible e.g. two socket outlet with each 22 kW, four socket outlets with each 11 kW or one socket outlet with 43 kW

Table 4 Construction costs for selected EV charging stations at motorway service areas

Manufacturer	Model	A	USD (VAT not included[a])			
			Charging station	Connection[b]	Construction work[b]	SUM
RWE	eStationCITY SMART	63	7,091	5,670	7,000	19,824
Evergreen	Terra	63	8,818	5,670	7,000	21,551
Evergreen	Magnum	63	13,129	5,670	7,000	25,862
ABB	Terra SC Duo Charge	35	15,000	3,150	7,000	25,185
ABB	Terra 51	80	28,626	7,200	7,955	43,861
ABB	Terra 52	80	34,091	7,200	7,955	49,326
ABB	Terra 53 CJ	80	36,818	7,200	7,955	52,053

Sources Refs. [21–23]

Notes

[a] All prices are based upon construction of a series of charging station at a minimum of 20 different sites (motorway service areas)

[b] Budget items, unit prices and assumptions are shown in Table 10

9 Provision of Power to Charging Stations on the Motorway Network

In city areas, distribution system operators (DSO) charge a fixed rate for connection to the electricity grid based on the amount of ampere needed. The DSO is, however, entitled to give a discount for connection of EV charging stations to the grid. The ordinary rate is DKK 1,035/A (USD 188) and the discount rate is DKK 495/A (USD 90).

Motorways are considered to be located in rural areas. In rural areas, the DSO may charge the actual costs for connecting to the electricity grid.[10] The costs affiliated with providing the needed amperes for EV charging at the motorway network therefore depends on the available capacity at the site. If there is not sufficient excess capacity at the site, the costs for connecting to the electricity grid can be very large.

The business case below is based upon the discount rate for connection to the electricity grid. Although, at a number of motorway service stations, the costs for connecting to the electricity grid may be significantly higher.

10 Business Case for Establishing EV Recharging Station on the Danish Motorway Network

From a consumer perspective, it is important that there is a country wide network of high power recharging stations that allows the EV user to drive across the country. CLEVER's DC based recharging network consisted in January 2014 of 51 units of 50 kW DC chargers. The total number of EVs in Denmark that could use CHAdeMO and CCS DC charging solution at the same time amounted to 790.[11]

This corresponds to a frequency of DC charger per CHAdeMO/CCS ready EV of 0.065. Hence, there was one CHAdeMO-charging station for each 15.5 CHAdeMO/CCS ready EV. Compared to the demand for DC charging there was a significant overprovision of DC charging stations in Denmark.

Under assumption of 100 % occupancy rate of a charging station, one 50 kW CHAdeMO charging station can daily recharge 48 Nissan LEAFs from 0–80 % SOC.[12] If 5–10 % of the energy consumption needed to recharge EVs takes place at a CHAdeMO charging station, each station may cover the recharging needs of 440–880 CHAdeMO ready EVs (with a battery size of 24 kWh).

---

[10] Only relevant if the actual costs for connecting the site to the electricity grid exceeds the costs for purchasing amperes.

[11] Based on the Danish Car Importers Association's database of stock of EVs in Denmark.

[12] Charging a Nissan LEAF from 0–80 % SOC takes 30 min.

It is, however, not possible to achieve a 100 % occupancy rate. The EV driver uses time to drive forward to the charging station, connect the EV to the station, pay for the consumed energy and leave the station. To a limited extend there is a need for service and maintenance, which also reduces the possible occupancy rate.

It is therefore assumed that the highest possible occupancy rate of a charging station is 80 %. This corresponds to one charging station covering the DC charging needs of 350–700 EVs.

From this perspective, a significant growth in the number of CHAdeMO ready EVs on Danish roads is needed before the network of CHAdeMO recharging stations will need to be expanded.

As outline above, from the end of 2013 almost all EVs that are sold in Denmark will be equipped with a high power charging solution including 22/43 kW AC, 40–50 kW DC CCS and 50 kW DC. Presently, none of these technological trajectories will dominate the market. To make it convenient and attractive for EV drivers to use high power charging stations at a Danish motorway charging network, the charging stations have to support Type 2 (22 kW AC), CCS (50 kW DC) and CHAdeMO (50 kW DC).

By including all three standardized recharging systems, the consumer does not have to worry whether the EV is compatible with the available high power charging solution. Compatibility with all three standards also contributes to an increase in the customer base for the recharging provider. Lack of compatibility with the motorway charging system may on the other hand have a huge impact on which EV models that the consumers are willing to purchase.

A charging station that supports three standardized charging systems is in the following referred to as a *charging service station*. A charging service station may consist of one or more charging posts.

At the time of writing it was not possible to access to price data for a charging station that supports all three standards. In the business case below, the charging service station is made up by a combined 50 kW CHAdeMO and 50 kW CCS charging station and a separate 22 kW AC post. The charging service station is assumed to be connected to the electricity grid with 160 A. 80 A is allocated to DC charging and 63 A to AC charging.[13] The CHAdeMO/CCS station can recharge one EV at a time with peak load.[14] The 22 kW AC station can simultaneously charge two EVs.

---

[13] The purchase of ampere from DSOs is divided into the following steps (which also are the levels that fuses are available in): 25, 35, 50, 63, 80, 100, 125, 160, 200, 225 and 250 A. In larger amounts exact quantities of ampere can be purchased e.g. 440 A.

[14] 50 kW DC charging of an EV with a 24 kWh battery from 0–80 % SOC is assumed to take 30 min. AC 22 kW charging of an EV with a 24 kW battery from 0–80 % SOC is assumed to take 60 min.

The business case is based on the assumption that CLEVER's existing network of CHAdeMO charging stations are converted into a CHAdeMO/CCS charging station combined with a 22 kW AC station with 2 socket outlets.[15]

Further, it is assumed that 20 new charging service stations are established on the Danish motorway network. In 2014, the total number charging service stations in Denmark amounts to 70.

With 80 % occupancy rate, one recharging service station would daily be able to recharge 77 EVs.

The following calculations are based upon two scenarios:

- In the low scenario, 5 % of an EV's charging needs are covered with high power charging service stations. In this scenario each charging service station can cover up to 1,300 EVs' high power charging requirements.
- In the high scenario, 10 % of an EV's charging needs are covered with high power recharging service stations. In this scenario each charging service station can cover up to 650 EVs' high power charging requirements.

Expansion of the network of charging service station will happen as the Charging Point Operators experience a critical load of their stations. It is assumed that such a critical load is achieved when 22 EVs daily charge the battery from 0 to 80 % SOC. This assumption is based upon a very crude model for EV driver behavior, as the model does not include the impact of congestion and rush hour on consumer charging behavior.

Critical load is reached with 433 EVs in the low scenario and 217 EVs in the high scenario.

It is assumed that 75 % of all new EVs will use the network of charging service stations. A number of EVs in public and private fleets will probably never—or almost never—use charging service stations. Today, most EVs in Denmark belong to municipalities and companies. It is, however, assumed that in the future a high share of EVs will belong to private users that frequently use charging service stations.

Figure 8 reflects the need for expansion of the charging service station network at a critical load of 217 EVs per recharging service station (High scenario). The stock of EVs is based on The Danish Energy Association, Energinet.dk and Dong Energy's projection.

The calculation of EVs with high power recharging option and the expansion of the charging service station network is shown in Table 5.

Figure 9 illustrates the need for expansion of the charging service station network at a critical load of 433 EVs per charging service station (Low scenario).

CLEVER provides two different products that give EV drivers access to CLEVER's network of publicly available charging stations: "Go" and "Go More".

---

[15] CLEVER plans to build a number of the combined CCS and 22 kW AC charging station at new sites. A number of existing CHAdeMO charging stations will not be upgraded to combined CHAdeMO, CCS and 22 kW AC charging service station.

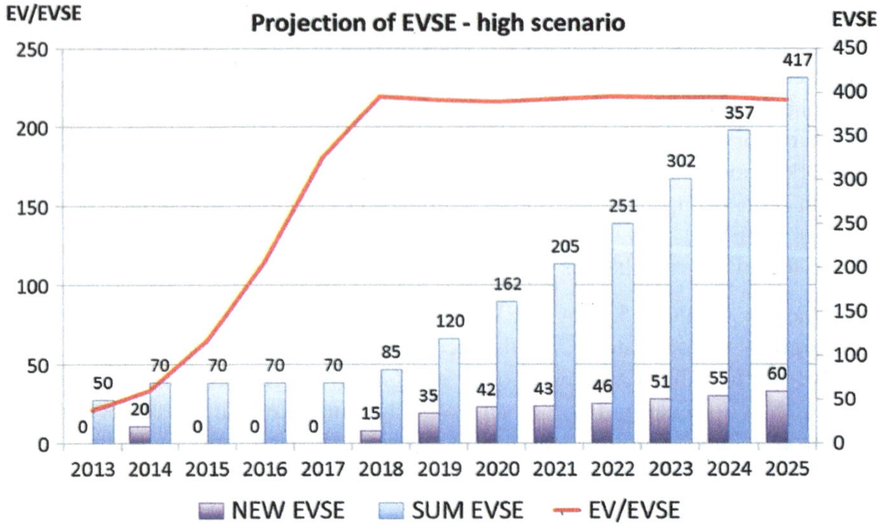

Fig. 8 Projection of high power recharging service stations (EVSE) and ratio of EVs/EVSE—High scenario. *NEW EVSE* New charging service stations; *SUM EVSE* Units of charging service stations in operation; *EV/EVSE* Units of EVs with high power recharging option/recharging service stations

Table 5 Projection of stock of EVs and high power recharging service stations

Year	Stock all EVs	EVs using high power recharging	High scenario			Low scenario		
			New EVSE	Sum EVSE	EV/ EVSE	New EVSE	Sum EVSE	EV/ EVSE
2013	1,533	1,043	0	50	21	0	50	21
2014	3,076	2,307	20	70	33	20	70	33
2015	6,047	4,535	0	70	65	0	70	65
2016	10,602	7,952	0	70	114	0	70	114
2017	16,842	12,632	0	70	180	0	70	180
2018	24,864	18,648	15	85	219	0	70	266
2019	34,768	26,076	35	120	217	0	70	373
2020	46,653	34,990	42	162	216	10	80	437
2021	59,529	44,647	43	205	218	23	103	433
2022	73,395	55,046	46	251	219	24	127	433
2023	88,249	66,187	51	302	219	25	152	435
2024	104,094	78,071	55	357	219	28	180	434
2025	120,927	90,695	60	417	217	28	208	436

With "Go" the customer only pays for the actual energy consumption. The price for "Go" customers is DKK 4.4/kWh (USD 0.8) (VAT not included). The price for "Go More" customers is DKK 2.8/kWh (USD 0.51) (VAT not included) and a monthly subscription fee of DKK 79 (USD 14.4) (VAT not included) [15].

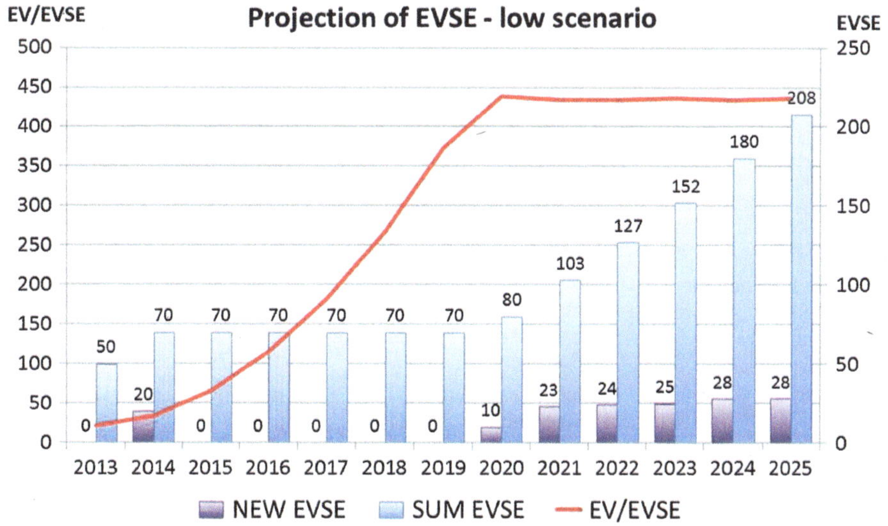

Fig. 9 Projection of high power recharging service stations (EVSE) and ratio of EVs/EVSE—Low scenario. *NEW EVSE* New charging service stations; *SUM EVSE* Units of charging service stations in operation; *EV/EVSE* Units of EVs with high power recharging option/recharging service stations

The business case is calculated for two cases where the Charging Point Operator gets a concession of 5 and 10 years respectively. For both cases the business case is calculated for the low and the high scenario.

The business case is based upon the following assumptions:

- An interest rate of 5 %.
- The cost per ampere is DKK 495/A (USD 90/A).
- Depreciation over 5 and 10 years respectively for the charging station equipment and all installation costs.
- In the low scenario, the energy consumption per EV for high power charging is 150 kWh/year.
- In the high scenario, the energy consumption per EV for high power charging is 300 kWh/year.
- A charging service station includes the installation of 1 unit ABB Terra 53 CJ (50 kW DC CHAdeMO/CSS) and 1 unit RWE eStationCITY SMART. The selected recharging stations have an expected lifetime of minimum 10 years [22, 24].
- Service agreement for a charging service station is expected to cost DKK 7,000/year (USD 1,273). The service agreement includes performance monitoring and debugging (online and on location).
- Construction costs as outlined in Table 4.
- A prerequisite for the applied price data is construction of charging service stations at a minimum of 20 sites.
- Billing costs are not included (will be passed to the consumer).

Table 6 Revenue for operating a charging service station for 5 years.

5 years concession	High scenario	Low scenario
Balance	46,228	−7,714
Including return on investment	30,573	−14,538

USD
Cost for connecting to the grid is USD 90/A

Table 7 Revenue for operating a charging service station for 10 years

10 years concession	High scenario	Low scenario
Balance	243,779	143,174
Including return on investment	163,466	113,668

USD
Cost for connecting to the grid is USD 90/A

Table 8 Revenue for operating a recharging service station for 5 years

5 years concession	High scenario	Low scenario
Balance	7.783	−46.159
Including return of investment	−6.042	−51.152

USD
Cost for connecting to the grid is USD 50,000

- Electricity price is set at DKK 0.87/kWh[16] (USD 0.16).
- The customer pays DKK 4.4/kWh (USD 0.8) (VAT not included). None subscription fee is charged.
- DC charging is used at 2/3 of all charging sessions at the charging service station, and the energy loss associated with converting AC to DC is estimated to be 4 % of the consumption.
- 75 % of all EVs are expected to use the charging service solutions.
- No fee is paid for using the Danish Road Directorate's road areas at the service stations.

The revenue from the business case is shown in Tables 6 and 7.

The business case has also been calculated for a case where the charging point operator has to pay the actual costs for connecting to the electricity grid—and hence not only USD 90/A. It is assumed that the actual costs amount to USD 50,000 (the costs may actually be even higher). Based on this assumption, a concession period of 10 years is needed for the investment to be attractive (see Tables 8 and 9).

[16] Based upon Eurostat data for small industrial energy consumers in Denmark.

Table 9 Revenue for operating a recharging service station for 10 years

10 years concession	High scenario	Low scenario
Balance	208,179	107,574
Including return of investment	129,561	79,763

USD
Cost for connecting to the grid is USD 50,000

11 Conclusion

The business case shows that construction of a network of charging service stations is commercially viable in the long-term—particularly from 2018–2020 where the occupancy rate of the charging stations is expected to reach a critical load.

If permission to operation of recharge service stations at the Danish motorway network only is granted for a period of 5 years, the risks associated with establishing the recharging stations is too high. This renders investments unattractive as the return rate on investments until 2018 will be too low.

There are significant uncertainties connected to some of the key assumption in the business model. If the market uptake of EVs turns out to be lower than anticipated or fewer EVs use the motorway charging stations, the income from the charging service stations may be significantly lower. There are therefore considerable risks associated with investments in EV charging service stations.

There are today large differences in the occupancy rate of existing DC charging stations in Denmark. Charging service stations at the motorway network are expected to have a relatively high occupancy rate, as the accessibility for EVs will be high.

EV charging typically takes longer than to the fill up an internal combustion engine car with gasoline or diesel. If the charging service station is situated next to a gas station kiosk or a restaurant, the EV drivers may use the break to buy more food and other products than the average car drivers. Revenue from such income is not included in the business case.

Appendix

Table 10 Unit costs for construction of EV high power charging stations at motorway service areas

Budget items	35 A	63 A	80 A	125 A	160 A	200 A	250 A
Grid connection fee for the first 25 A	2,250	2,250	2,250	2,250	2,250	2,250	2,250
Grid connection fee for A > 25 A (USD/A)	90	90	90	90	90	90	90
Excavation non-paved area, piping, wiring and restoration (USD/per meter)[a]	91	91	100	109	118	127	145
Excavation paved area, piping, wiring and restoration with asphalt (USD/per meter)[a]	164	164	173	182	191	200	218
Installation of electrical panel and meter	3,091	3,091	3,636	4,182	4,545	5,091	5,455
Construction of concrete foundation for EV charging station	909	909	1,091	1,091	1,273	1,273	1,273

USD (VAT not included)
Sources GodEnergi [25]
Note
[a] Possible requirements from the Danish Road Directorate for signposting are not included. It is assumed that the soil is not contaminated

References

1. European Commission (2013) Directive of the European Parliament and of the council on the deployment of alternative fuels infrastructure, COM(2013) 18 final, Jan 2013
2. Danish Transport Ministry (2013) Press announcement September 4th 2013. http://www.trm.dk/da/nyheder/2013/udbud+skal+sikre+ladestandere+til+elbiler+p%C3%A5+statsvejnettet/
3. Victor Hug Consult (2013) Redegørelse for teknologier til opladning af elbiler ved motorvejsnettet, prepared by Victor Hug Consult for the Danish Energy Agency, July 2013 (Not Published)
4. CEM EVI, IEA (2013) Global EV Outlook. Understanding the electric vehicle landscape to 2020, Clean Energy Ministerial EV Initiative and International Energy Agency, April 2013
5. DEA, DTA (2013) Energistyrelsen og Trafikstyrelsens statistik for ladeinfrastruktur i Danmark (the Danish Energy Agency and the Danish Transport Authorities statistics of EVSE in Denmark), data collected by Victor Hug Consult for the Danish Energy Agency and the Danish Transport Authorities, Nov 2013
6. CLEVER (2014) Press announcement January 31st 2014. https://www.clever.dk/nyheder/her-opsaetter-vi-combo-hurtiglade-stationer-i-foraaret/

7. Victor Hug Consult (2013) Energistyrelsens markedsanalyse af elbiler på det danske marked 2013-14, prepared by Victor Hug Consult for the Danish Energy Agency, Oct 2013
8. SMC (2013) Personal communication. Jonas Wallin, Skandinavisk Motor Company, 23rd Oct 2013
9. Messer Thomsen M (2013) ZOE elbil ladning med 22 kW, June 2013. http://evtest.dk/zoe-elbil-ladning-med-22-kw/
10. ABB (2012) Example charge sessions MiEV and leaf. Laboratory test conducted by Johan Kaptein, ABB Feb 2012
11. Nissan (2013) Den nye Nissan Leaf - bygget i Europa til Europa. http://www.newsroom.nissan-europe.com/dk/da-dk/Media/Media.aspx?mediaid=103300
12. Renault (2013) Technical Specification Renault Zoe
13. DEA (2012) Statusrapport for Energistyrelsens forsøgsordning for elbiler (Status report for the Danish Energy Agency's EV Test Scheme), prepared by the Danish Energy Agency and Victor Hug Consult, Aug 2012 (Not Published)
14. CLEVER (2013) Test-en-elbil kvartalsrapport 2. kvartal 2013/6, CLEVER, July 2012
15. CLEVER (2013) https://www.clever.dk/produkter/opladning-paa-farten/
16. ABB (2013) Personal communication, Per Rømer Kofod, ABB, 3 May 2013
17. EN.dk et al (2013) Scenarier for udrulning af elbiler i Danmark, Energinet.dk, Danish Energy Association and Dong Energy 2013
18. EN.dk (2013) Correspondence, Christoffer Nicolaj Rasch, Energinet.dk, 14 May 2013
19. CleanCharge (2013) https://www.cleancharge.dk/produkter
20. ABB (2013) www.abb.com/products/us/9AAC172688.aspx
21. Evergreen (2013) Correspondence, Lars Bøegh Nielsen, Evergreen, 26 June 2013
22. ABB (2013) Correspondence, Per Rømer Kofod, ABB, 31 May 2013
23. CleanCharge (2013) Correspondence, Nils Dullum, CleanCharge, 29 May 2013
24. CleanCharge (2013) Personal communication, Nils Dullum, CleanCharge, 30 June 2013
25. GodEnergi (2014) Correspondence, Jan Darville, GodEnergi, 29 Jan 2014

Pricing Plug-in Electric Vehicle Recharging in Multi-unit Dwellings: Financial Viability and Fueling Costs

Brett Williams and J.R. DeShazo

Abstract This research explores whether pricing structures and levels likely to provide electric vehicle drivers with financial motivation to recharge at multi-unit dwellings might provide sufficient opportunity for station cost recovery. Compared to a popular 50 mpg gasoline hybrid baseline, residential charging prices might have to be kept below \$0.26/kWh, \$1.00/h of charging, or \$85/month—levels that only support roughly \$1,000–2,000 in facility investment per vehicle served. Increasing facility utilization while minimizing per-vehicle costs is key to improving financial viability and, across pricing structures, could more than double the cost recovery potential. Further, site hosts' choice of pricing structure will differentially affect their ability to remain financially viable in the face of input-parameter uncertainty.

Keywords Plug-in electric vehicle · Recharging · Multi-unit dwellings (MUDs) · Pricing · Recharging costs · Profitability · Cost recovery · Revenues · Financial viability

1 Introduction

1.1 Background, Objectives, and Article Structure

Overnight recharging at home is expected to be the most prevalent and cost-effective way to refuel plug-in electric vehicles (PEVs) [1]. This is due to factors such as long and regular vehicle residence times [2]. Nevertheless, the implementation of

B. Williams (✉)
Center for Sustainable Energy, 9325 Sky Park Court #100, San Diego, CA 92123, USA
e-mail: brett.williams@energycenter.org

J.R. DeShazo
Luskin Center for Innovation, University of California at Los Angeles,
337 Charles E Young Dr. East, Los Angeles, CA 90095, USA
e-mail: deshazo@ucla.edu

D. Beeton and G. Meyer (eds.), *Electric Vehicle Business Models*,
Lecture Notes in Mobility, DOI 10.1007/978-3-319-12244-1_6

residential recharging remains particularly challenging in multi-unit dwelling (MUD) environments as stakeholders wrestle with this relatively new phenomenon and its potential costs and benefits in a wide variety of contexts. Site hosts (e.g. landlords or property owners), regulators, and consumers lack information needed in order to understand the costs of fueling and implications of various pricing and incentive policies. The supporting literature is relatively sparse and nascent. For example, business models for recharging have been discussed from the perspective of market structure and actors [3, 4]. Financial analysis by Schroeder and Traber examined public fast charging stations [5], which use different technology with different considerations than examined here. Botsford [6] analyzed non-residential charging from a cost-based perspective that articulates what revenues are required to cover examined costs, the opposite approach to the one taken below for MUDs.

This research assesses recharging at MUDs from two main perspectives: (i) site hosts investing in MUD recharging facilities and pricing their use and (ii) resident PEV drivers. These perspectives are explored in turn in each of Sect. 2 (methods and assumptions) and Sect. 3 (results and discussion). This analysis makes several contributions, including:

1. exploring the opportunity for facility cost recovery at prices that resident drivers might find financially motivating,
2. describing opportunities for increasing financial viability through economies of scale in use,
3. characterizing each of three pricing structures for their differential impacts on drivers with varying driving and vehicle characteristics,
4. describing how choice of pricing strategy affects facility viability in the face of uncertainty, and
5. providing benchmarks that facilitate comparison of pricing levels both across pricing structures and relative to two gasoline refueling baselines.

2 Methods and Assumptions

The following describes the framework and assumptions used to analyze (1) MUD recharging facility financial viability (Sect. 2.1), and (2) fueling costs for PEV drivers in MUDs (Sect. 2.2).

2.1 MUD Recharging Facility Financial Model Elements

This section describes the major elements of the financial model developed to examine recharging investments from the site-host perspective, including costs, financial assumptions, and facility utilization.

2.1.1 Costs

Recharging-station costs can be broken into three types: upfront, periodic, and variable costs. Costs often vary dramatically based on site-specific conditions, and not all costs are required for all installations.

Upfront costs include the fully-burdened cost of the facility and its installation, including:

1. PEV-ready electrical service
   (e.g., site assessment and design, electric-service upgrades, permitting, trenching, conduit);
2. parking/"station" modifications
   (e.g., accessways, bulwarks, signage, security, access control, data logging if separate from the charger);
3. electric-vehicle supply equipment (EVSE)
   (e.g., chargers with various configurations of power level, number of outlets or vehicles served, cabling, access controls, network access capability, data logging); and
4. facility decommissioning

Variable costs (e.g., electricity energy and demand charges, rate-tier adjustments, sales tax, facility operation and maintenance) relate to the amount of charging provided (e.g., per kilowatt-hour [kWh]). Periodic costs are ongoing but relate less closely to the amount of service provided (e.g., property tax, insurance, periodic access or network fees, facility management and data processing). They can be treated as an additional upfront, fixed lump sum if their level is known.

Fixed Costs (Upfront and Periodic) Because of the wide variety of facility cost structures (reserved for future work) and in order to allow flexible exploration of a variety of cost levels, the financial modelling employed herein does not attempt to model MUD recharging facility costs. Rather, it explicitly presents a range of "all-in" fixed investment levels (one per row in Tables 1 and 3). This allows the reader to choose different levels (i.e., pick different rows) appropriate to different situations, as described in Sect. 3.1.

Variable Costs Within a reasonable range of utilization, electricity costs are expected to dominate the variable cost category. Electricity costs vary based on utility territory, customer class, total energy and power demanded, season of year, time of day, and rate schedule selected. For simplicity, it is assumed that variable costs average $0.1640/kWh, the average price of residential electricity in California[1] in the most recent quarter (2Q2013) [7].

---

[1] This assumption does not fully take into account the possibility that energy purchased to charge PEVs could move the building into a more expensive tier of electricity prices and/or that the power demanded by the equipment could add to the facilities' demand charges (if applicable). This is probably a reasonable simplification for a small number of vehicles served relative to the location's overall electricity consumption and with a little care not to allow PEV charging during the hours each month when the facility demands its peak amount of power (upon when demand charges are set).

Table 1 MUD recharging investment 10-year net present value: pricing scenarios[a]

(a) Fee structure per-kWh Session fee $0.00

Electricity markup

Project Cost	$ -	$ 0.10	$ 0.20	$0.30
$ -	$ (0)	$ 2,763	$ 5,526	$ 8,289
$ 1,000	$ (1,437)	$ 1,326	$ 4,089	$ 6,852
$ 2,000	$ (2,875)	$ (112)	$ 2,652	$ 5,415
$ 3,000	$ (4,312)	$ (1,549)	$ 1,214	$ 3,977
$ 4,000	$ (5,750)	$ (2,986)	$ (223)	$ 2,540
$ 5,000	$ (7,187)	$ (4,424)	$ (1,661)	$ 1,103
$ 6,000	$ (8,624)	$ (5,861)	$ (3,098)	$ (335)
$ 7,000	$ (10,062)	$ (7,299)	$ (4,535)	$ (1,772)
$ 8,000	$ (11,499)	$ (8,736)	$ (5,973)	$ (3,210)

(b) Fee structure per-kWh Session fee $1.00

Electricity markup

Project Cost	$ -	$ 0.10	$ 0.20	$ 0.30
$ -	$ 2,703	$ 5,466	$ 8,229	$10,992
$ 1,000	$ 1,265	$ 4,028	$ 6,792	$ 9,555
$ 2,000	$ (172)	$ 2,591	$ 5,354	$ 8,117
$ 3,000	$ (1,610)	$ 1,154	$ 3,917	$ 6,680
$ 4,000	$ (3,047)	$ (284)	$ 2,479	$ 5,243
$ 5,000	$ (4,484)	$ (1,721)	$ 1,042	$ 3,805
$ 6,000	$ (5,922)	$ (3,159)	$ (395)	$ 2,368
$ 7,000	$ (7,359)	$ (4,596)	$ (1,833)	$ 930
$ 8,000	$ (8,796)	$ (6,033)	$ (3,270)	$ (507)

(c) Fee structure per-hour Session fee $0.00

Hourly fee

Project cost	$ 0.65	$ 1.00	$ 1.35	$ 1.70
$ -	$ (2)	$ 2,761	$ 5,524	$ 8,287
$ 1,000	$ (1,439)	$ 1,324	$ 4,087	$ 6,850
$ 2,000	$ (2,877)	$ (114)	$ 2,649	$ 5,413
$ 3,000	$ (4,314)	$ (1,551)	$ 1,212	$ 3,975
$ 4,000	$ (5,752)	$ (2,988)	$ (225)	$ 2,538
$ 5,000	$ (7,189)	$ (4,426)	$ (1,663)	$ 1,101
$ 6,000	$ (8,626)	$ (5,863)	$ (3,100)	$ (337)
$ 7,000	$ (10,064)	$ (7,301)	$ (4,537)	$ (1,774)
$ 8,000	$ (11,501)	$ (8,738)	$ (5,975)	$ (3,212)

(d) Fee structure per-hour Session fee $1.00

Hourly fee

Project cost	$ 0.65	$ 1.00	$ 1.35	$ 1.70
$ -	$ 2,701	$ 5,464	$ 8,227	$10,990
$ 1,000	$ 1,263	$ 4,026	$ 6,789	$ 9,553
$ 2,000	$ (174)	$ 2,589	$ 5,352	$ 8,115
$ 3,000	$ (1,612)	$ 1,152	$ 3,915	$ 6,678
$ 4,000	$ (3,049)	$ (286)	$ 2,477	$ 5,240
$ 5,000	$ (4,486)	$ (1,723)	$ 1,040	$ 3,803
$ 6,000	$ (5,924)	$ (3,161)	$ (397)	$ 2,366
$ 7,000	$ (7,361)	$ (4,598)	$ (1,835)	$ 928
$ 8,000	$ (8,798)	$ (6,035)	$ (3,272)	$ (509)

(e) Fee structure per-month Electricity fee $0.00

Monthly fee

Project cost	$ 55	$ 85	$ 115	$ 145
$ -	$ (37)	$ 2,743	$ 5,522	$ 8,302
$ 1,000	$ (1,475)	$ 1,305	$ 4,085	$ 6,865
$ 2,000	$ (2,912)	$ (132)	$ 2,648	$ 5,427
$ 3,000	$ (4,349)	$ (1,570)	$ 1,210	$ 3,990
$ 4,000	$ (5,787)	$ (3,007)	$ (227)	$ 2,553
$ 5,000	$ (7,224)	$ (4,444)	$ (1,664)	$ 1,115
$ 6,000	$ (8,662)	$ (5,882)	$ (3,102)	$ (322)
$ 7,000	$ (10,099)	$ (7,319)	$ (4,539)	$ (1,759)
$ 8,000	$ (11,536)	$ (8,756)	$ (5,977)	$ (3,197)

(f) Fee structure per-month Electricity fee $0.1640 (year 1)

Monthly fee

Project cost	$ 55	$ 85	$ 115	$ 145
$ -	$ 5,096	$ 7,876	$10,656	$13,436
$ 1,000	$ 3,659	$ 6,439	$ 9,219	$11,998
$ 2,000	$ 2,222	$ 5,001	$ 7,781	$10,561
$ 3,000	$ 784	$ 3,564	$ 6,344	$ 9,124
$ 4,000	$ (653)	$ 2,127	$ 4,906	$ 7,686
$ 5,000	$ (2,091)	$ 689	$ 3,469	$ 6,249
$ 6,000	$ (3,528)	$ (748)	$ 2,032	$ 4,812
$ 7,000	$ (4,965)	$ (2,185)	$ 594	$ 3,374
$ 8,000	$ (6,403)	$ (3,623)	$ (843)	$ 1,937

[a] Assumes a 5 % discount rate, 350 commute days, 30 miles of daily driving, 10.2 kWh consumed, 3.5-kW charging (L2), 2.9-h/session, and $0.1640/kWh electricity costs in year 1. Electricity and maintenance costs are escalated by 3 % per year

For the purposes of uncertainty analysis, it is assumed variable costs range from a U.S. low of $0.0867/kWh (2Q2013 Washington) to a high of $0.3704/kWh (2Q2013 Hawai'i) [7].

2.1.2 Financial Assumptions

Unless otherwise stated, the financial modelling described here evaluates the present value of charging revenues net of all-in investment costs assuming: a 10-year planning horizon, a 5 % discount rate, a 3 % annual growth rate in electricity and gasoline prices and the level of markup, and maintenance costs

---

(Footnote 1 continued)
Further, utilizing special EV rates (possibly requiring the purchase of a second meter to be included in "all-in" installation costs) may also avoid these possibilities.

equivalent to 5 % of total costs. Taxes and revenue sharing with network providers are not treated explicitly here, and thus can either be considered to be 0 % or covered in fully-burdened fixed costs.

2.1.3 Facility Utilization Assumptions

Variable costs and revenues depend on the level of use experienced by the charging facility. Electricity costs and per-kWh revenues depend on the amount of energy consumed, and per-hour revenues depend on the charging duration. These are in turn a function of the power (kW) of the charging equipment, and the amount of energy (kWh) required. For simplicity, several assumptions are made, including that the vehicle will draw power until it is fully charged, that the power drawn is constant, and that it amounts to approximately 3.5 kW for Level 2 charging [8]. The amount of energy required is dependent on the state of charge of the vehicle when it plugs in, which depends largely on the daily driving distance, assumed here to average 30 miles per day [8].

In order to calculate how many kWh will be needed to recharge PEVs that have traveled, on average, that daily commute distance (30 miles), the vehicle's electric fuel economy is needed. It is assumed that the average PEV can make the trip in electric mode consuming electricity at approximately 34.1 kWh/100 miles. This is an average of the U.S. Environmental Protection Agency's adjusted electric economy ratings for PEVs weighted by aggregated sales data through August 2013 [9].

Thus, to recover from a 30-mile day, the charging facility needs to provide 0.341 kWh/miles or 10.2 kWh per vehicle. At the 3.5-kW Level 2 charging rate, this would take approximately 2.9 h. It is further assumed that the charging equipment will be utilized 7 days per week for 50 weeks out of the year, or 350 days per year.

2.2 Fueling Costs Calculations for Resident Drivers: Additional Inputs

This section describes additional assumptions used in calculating the costs of recharging at home for MUD residents facing the variety of pricing structures and levels described above. Key additional inputs used include (1) the sales-weighted, EPA-rating plug-in-hybrid average gasoline fuel economy of approximately 41.1 mpg (miles/gal) and (2) conventional-vehicle fuel economy of 27.2 miles/gal. The former was calculated based on PEV sales [9] and the latter is the EPA composite rating for small and medium cars in model year (MY) 2011, the most recent year for which data was available [10]. This is a higher, and therefore more conservative, fuel economy than the composite ratings for both MY 2011 cars as a whole (25.9) and all passenger vehicles (including trucks, 22.8). At the moment, PEVs are mostly small and medium cars, but the alternative fueling option available to some PEV drivers might be a larger car or truck. In all cases (including PEVs), the ratings used herein are EPA "adjusted" to better reflect real-world driving conditions (i.e., the number used on the new-vehicle sticker).

3 Results and Discussion

3.1 MUD Recharging Facility Financial Viability and Pricing Options

This section presents an analysis of recharging station profitability as a function of various pricing structures and levels (Sect. 3.1.1) and examines both uncertainty in input parameters (Sect. 3.1.2) and increasing station utilization (Sect. 3.1.3).

3.1.1 Residential Recharging Facility Financial Viability

Table 1 illustrates the effect that various inputs, including nine "all-in" investment cost levels (one per row), have on the present value of recharging facility net revenues. In situations where the net present value is positive, costs are recovered and the facility investment is potentially profitable. The table allows exploration of individual situations seen by MUDs at specific locations with varying conditions. Additionally, it allows exploration of the effect of incentives that change cost levels, reserved for future work.

Table 1 has six parts (a–f) presenting the 10-year present value of net revenues[2] resulting from three basic price structures: per-kWh (a, b), per-hour (c, d), and per-month (e, f). Per-kWh and per-hour structures are presented both with (b, d) and without (a, c) an additional fixed fee per charging session. The per-month structure is presented both with an additional fee to cover electricity costs (f) and without the additional fee (e).

Each column of the table is for a different fee (price) level. Per-kWh scenarios include columns based on the amount of the markup added to the electricity costs

[2] It should be noted that the only revenues represented in Table 1 are those from fees for recharging services. Other sources of revenue or broader benefits might be available, including from tax and accounting benefits, participation in utility demand-response programs, and future value streams from the intelligent control of charging rates to provide various types of grid services (e.g., participation in regional grid markets like regulation, benefits to utility operation and the transmission or distribution system, customer-side-of-the-meter benefits like utility-bill mitigation or power quality/reliability, and/or a variety of related renewable-integration services [11, 12]). Eventually, recharging systems might be upgraded to broker bi-directional power flows to and from PEVs for greater levels of grid services, onsite energy management, and emergency power.

passed on to the driver. Per-hour and per-month scenarios include columns based on the level of those fees. The first fee-level column in parts a, c, and e presents the approximate break-even level required to cover variable (electricity) costs only (i.e., zero project costs). For example, in part a, charging for the electricity with no additional markup precisely covers electricity costs, resulting in a net present value of zero. $0.65/h and $55/month are rough equivalents, given electricity costing $0.1640/kWh, 30 miles, and 350-day-per-year use. The next three columns represent symmetrical increases in the fee level for illustration up to levels that might represent reasonable maximums that drivers are generally willing to pay. As will be seen in Sect. 3.2, the third column might still provide the driver with some advantage over fueling on gasoline at today's prices, whereas the fourth column might be considered uncompetitive with gasoline, on the whole. The exception are the fees in part f: to achieve similar cost-recovery potential, higher fees (per-month + electricity charges) are required and will be less active to driver than the equivalent fees shown in parts b and d.

In general, it appears possible to recover roughly $5,000 of investment on one charging unit. This is not unlike the initial situation facing many locations that:

1. want to test the waters by providing one charging point, or whose cost structure makes it difficult to provide the first few parking spots with EVSE at costs much less than a couple thousand dollars per unit;
2. wish to limit charging in the afternoon and evening to avoid peaks in either energy charges or facility demand charges (if applicable);
3. do not wish to create a potentially complex, costly, and/or driver-time-consuming system involving moving of cars in and out of charging locations to increase the number of charge events per day; and/or
4. do not have synergistic opportunities to open their recharging facilities to public vehicles when resident charging is unneeded.

Nevertheless, it is clear that, for those locations concerned with profitable operation of their recharging facilities, increasing utilization and/or reducing average unit costs are important. These topics are explored in Sect. 3.1.2 and future work.

Sub-tables b and d show the effect of adding an additional $1 fixed fee per session (for per-kWh and per-hour structures). This increases the cost-recovery potential of a given simple fee structure from up to $5,000–6,000 to up to $7,000–8,000 in investment. The effect is similar for the per-month structure (not shown), and somewhat smaller than charging an additional fee to cover electricity costs in the per-month structure, which equates in this driving scenario to a $1.68 fixed fee (sub-table f). In a similar vein, the site host might be tempted to confound parking and recharging pricing in such a way that the PEV driver continues to get charged for occupying a recharging space after active recharging is completed. For a given amount of time after charging, these parking fees act as additional fixed (relative to the amount or duration of charging) fees.

However, for reasons discussed here and elsewhere [13], fixed-fee structures are potentially both less transparent and more discriminatory against certain vehicle types and drivers. For example, for those with smaller batteries and/or shorter commutes, the large fixed component of these fee structures can, in one manifestation or another, quickly end up raising the effective costs of recharging to several dollars, not several cents, per kilowatt-hour received. This might be counterproductive to adoption of MUD charging by the majority of the PEV market that drives plug-in-hybrids or neighborhood electric vehicles (NEVs).

3.1.2 Sensitivity and Uncertainty Analysis of Financial Viability

Understanding the effects of uncertainty on financial viability is important to evaluate the robustness of net-present-value estimates. Additionally, if pricing structures respond differently to sources of uncertainty (e.g., in the daily driving distance of residents, maintenance and electricity costs, etc.), site hosts may be able to minimize variability in financial returns through their choice of pricing structure. This section explores uncertainties and sensitivities first using the $0.30-markup/kWh price structure and level as the "base case" scenario. Following the base-case explanation, the roughly equivalent[3] $1.70/h and $145/month fee structures are also analyzed. The results are summarized in Table 2.

(**a**) **Per-kWh Case** The net-present value (NPV) of the scenario in the third cost row and last markup column in Table 1a is estimated to be approximately $1,103. This indicates that a variable fee with a $0.30/kWh markup[4] is able to recover $5,000 in fixed project costs. Of the simple (i.e., without session fee) variable-fee scenarios discussed, this is the scenario able to cover the highest fixed project costs. It is taken as the per-kWh "base case," and its more general underlying assumptions are taken as "baseline" assumptions.

To explore the importance of various inputs to this base-case estimate of $1,103, a Monte Carlo simulation of 50,000 trials was run on the input parameters as described in Table 2 using Oracle's Crystal Ball software. The "best-guess" input assumptions discussed so far are in bold and have been bounded by ranges defined by "minimum" and "maximum" estimates based on a combination of the literature sources used to produce the corresponding point estimate and author judgment. All but one range have been characterized with "triangle" probability distributions defined by linearly decreasing probability from the "best-guess" to the minimum and maximum estimates. Maintenance cost escalation, for which it was thought particularly little was known about the appropriate probability distribution, was assigned a uniform distribution across the range of values considered.

By repeatedly re-calculating the $1,103 net-present value estimate using input values that are probabilistically picked from within the ranges described above, the

[3] Both roughly in terms of cost-recovery potential, but, more particularly, to the driver—see Sect. 3.2.

[4] i.e., ($0.1640 + $0.30)/kWh in year 1, where the electricity cost is escalated by 3 % per year.

Table 2 Uncertainty and importance of input parameters: per-kWh, per-hour and per-month

				+$0.30/kWh	$1.70/h	$145/mo.
Point estimate				$1,103	($1,101)	($91)
Monte Carlo mean				$946	($2,056)	($977)
95 % confidence interval				($2,879) to $5,493	($9,478) to $9577	($7,216) to $3,371
Input parameter	Min.	Best guess	Max.	Uncertainty Contribution (%)[a]		
Daily driving distance (miles)	15	30	45	64	4	−25
Maint. costs (% of all-in costs)	1 %	5 %	10 %	−13	−3	−9
PEV electric fuel economy (kWh/100 miles)	28.6	34.1	43.2	10	1	−4
Drive days per year	235	240	260	7	0.5	−3
Discount rate	3 %	5 %	10 %	−4		−1
Escalation of markup	1 %	3 %	5 %	3		
Maintenance cost escalation	1 %	Uniform (3 %)	5 %	−0.3	−0.1	−0.2
Charging power (kW)	1.4	3.5	7.2		−68	
Electricity cost (/kWh)	$0.0901	$0.1275	$0.30		−20	−52
Electricity cost escalation	1 %	3 %	12 %		3	−6

[a] Described in the text, this is a metric based on normalized rank correlation coefficients

Monte Carlo simulation produced a distribution of NPV estimates reflecting uncertainty in the input assumptions that can be compared to $1,103 point estimate (The latter, point estimate implicitly assumes perfect knowledge about input values).

In contrast to **the point-estimate of $1,103 for the NPV**, the Monte Carlo analysis produced a relatively symmetrical beta distribution with **a mean of $946 and a 95 % confidence interval** (C.I.) ranging **from ($2,879) to $5,493**.

This suggests that the NPV may be somewhat lower than the point estimate indicates, but that more room exists within the 95 % confidence interval for upside potential than downside potential. The last column in Table 2 gives the "contribution" produced by the specified uncertainty in each input parameter. This "contribution" is an illustrative metric produced by the Oracle software by normalizing the rank correlation coefficients between each input and the NPV estimate to illustrate how uncertainty in each input contributes to the overall distribution of NPV estimates produced over the course of the 50,000 estimations. Again, it is presented here simply to rank and roughly characterize the importance of each input and its uncertainty to the value of the NPV estimate of the base case.

The two dominant uncertainties were:

1. the daily driving distance (which determines how much electricity is needed and the range for which was kept wide to acknowledge drivers are distributed across a variety of commute distances and not stacked up near the average value); and
2. maintenance costs (currently modelled as a percentage of all-in fixed costs, which in turn are the most important factor in the NPV estimation but taken as given for each row of Table 1).

The electric fuel consumption of the vehicle (which also determines how much electricity is needed, but which is bounded more tightly by the range of current vehicles on the market) contributed just less than 10 %. The uncertainty in the number of drive days per year contributes about 7 % in this analysis as structured, and the discount rate and escalations contributed 4 % or less each. Parameters related to the variable cost (i.e., cost of electricity) do not contribute in this variable-rate pricing scenario, as expected.

This analysis suggests that maintenance costs need to be better understood and modelled. Additionally, if a wide range of potential drive distances need to be considered rather than a representative average, the results can be expected to vary considerably, making the decision-making process more complicated. More generally, a clearer picture is needed of facility utilization (see Sect. 3.1.2).

It should also be noted that the exact method of markup may be important and should be examined further. For example, would the markup be a fixed amount or a percentage? Or would the resident driver be asked to pay in terms of a specific markup or a specific total price/kWh consumed? How and how often would the markup or total change over time?

(b) Per-Hour As compared to the point NPV estimate of $1,101 for $1.70/h from Table 1, the Monte Carlo simulation produced a somewhat left-skewed max. Extreme distribution with a mean value of ($1,980) and a 95 % confidence interval of ($9,478)–$9,576. This indicates that $1.70/h may be much less likely to cover $5,000 in project costs than the point estimate indicates, and that there is considerable room for both downside and upside (though the median and mode are both more negative than the mean). Examining the contributions of the input parameters in Table 2, nearly 70 % is due to the possibility that higher charging power might decrease the active charging time and thus reduce billable hours. Uncertainty in electricity costs contributes another 20 %. Uncertainty in the daily driving distance contributes roughly 4 %. Similar in size but opposite in direction, the possibility that maintenance costs or increases in electricity prices might be worse than expected also show a modest potential to reduce cost recovery in this fee structure. Overall, utilization is not very important.

(c) Per-Month As compared to the point NPV estimate of $1,115 for $145/month from Table 1, the Monte Carlo simulation produced a somewhat right-skewed Weibull distribution with a mean value of ($977) and a 95 % confidence interval of ($7,216) to $3,371. This indicates that $145/month may be less likely to cover $5,000 in project costs than the point estimate indicates, and that there is much more room for downside than upside. Over half of the "contribution" is due to the possibility of higher electricity costs. The assumed daily driving distance contributes over a quarter, making it relatively important to this price structure. In this case, however, it is important to remember that decreased utilization is desirable from a cost-recovery standpoint. This creates a perverse incentive to collect monthly fees from resident drivers but discourage PEV charging and its associated benefits. Contributing less than 10 % are the possibility that maintenance costs might be worse than expected, higher-than expected electricity cost escalation, poorer than expected vehicle economy, and fewer drive days per year than expected.

Summary and Comparison of Uncertainty Across Fee Structures

Fundamentally, driving distance is important to the per-kWh structure, as it represents sales volume. Driving distance is also important to the per-month structure, but for the opposite reason (increased charging of PEVs decreases cost-recovery). Charging power, the rate or "speed" of charging, is critically important for per-hour viability, as it determines the active charging time and thus billable hours (for reasons discussed in Sect. 3.1). Electricity cost factors are important to both the per-hour and, in particular, per-month structures. Additionally, uncertainty in maintenance costs is important to all structures (though somewhat less so for the per-hour structure), flagging this as a priority for future refinement in the model.

Across fee structures, the effect of uncertainty in the input assumptions is to lower the NPV estimate but to provide significant room for upside potential. The per-kWh structure fares the best (smallest reduction in NPV estimate and large upside potential). The per-hour structure NPV estimate is lowered the most and has a large range (both upside and downside). The per-month upside potential is the most limited and the downside potential grows large if utilization and costs increase. The analysis of these inputs also indicates that site hosts wishing to minimize the variability in financial viability overall should adopt the per-kWh structure.

3.1.3 Revenue Scenarios: Increasing Utilization to Improve Cost Recovery

Thus far, the financial analysis has focused on the initial installation and use of one Level 2 charger by one PEV once per day, perhaps reasonable for MUDs with assigned parking. Under these conditions, and the various other assumptions described for the base case above, even the variable fee structure could only recover approximately $5,000 in project costs. Depending on the MUD environment (e.g., location of electrical panel relative to the parking, amount of required trenching), this may or may not be adequate for facility construction and installation. However, several opportunities exist to improve the picture. They include:

1. shuffling multiple resident vehicles through the EVSE-equipped parking spaces,
2. utilizing the same equipment for fleet and/or public charging when not in use by residents, ideally during off-peak times, and
3. installing low-cost, multiplexing[5] and/or low-power (e.g., Level 1) equipment.

The first two options may not be appropriate for certain MUDs. However, if available, they can be considered to improve the cost-recovery calculus in a straightforward way—for example by increasing the number of charge sessions per day or per year. The third, technological option may be the most widely desirable as it has the potential to increase utilization while simultaneously lowering certain project costs or barriers (e.g., electric panel capacity, equipment power ratings, etc.).

Regardless of the means by which utilization is increased, if equipment use increases significantly, the argument grows for either assessing at least part of the maintenance and operation costs on a variable basis per kWh consumed and/or explicitly accounting for accelerated equipment replacement, effects reserved for future work.

Table 3 illustrates the effect of increasing Level 2 equipment utilization on the NPV of the base case. It does so using three metrics: the average number of kWh charged per day, the average number of electric miles provided per day, and the number of PEVs served. The first two are directly related to one another, and the third (number of PEVs served) becomes important for the per-month price structure (highlighted below). It also explicitly examines one *decreased* utilization scenario (treated implicitly above as the lower bound in the uncertainty analysis), that where the vehicle returns home with only 15 miles of charge to recover (e.g., a shorter-than-average driving day, or one involving additional, non-residential charging).

As seen for the $0.30/kWh-markup case in Table 3a, doubling the utilization of the charger to 20.4 kWh per day (equivalent to 60 e-miles) yields enough revenue to support over $10,000 of investment. If the charger could be used in this way without undue valet, peak-electricity, and/or resident-inconvenience costs, it is likely to recover the investment required at a wide variety of MUDs. A very similar picture is seen for the roughly comparable $1.70/h price structure and level.

The per-month fee structure presents a more complex picture. Increasing facility utilization by a given number of customers *decreases* cost recovery potential as increasing electricity costs eat away at fixed subscription-fee revenues. This is seen when comparing the first and second columns in Table 3c, for 1 PEV, and when comparing the third and fourth columns, for 2 PEVs. However, comparing the second and third columns illustrates the effect of increasing the number of per-month subscribers from 1 to 2, thereby doubling revenues from $145 to $290 per month. Where multiple per-month subscribers might be able to use a single charger, the cost-recovery picture improves dramatically. As indicated above, this could be

[5] i.e., equipment that can charge multiple vehicles using one circuit and/or off-board charger, e.g., through use of multiple cords and control of the amount or timing of power sent to each vehicle.

Table 3 Utilization scenarios

(a) $0.30/kWh markup

Project cost	Utilization per day: 0.5 PEVs 15 e-mi 5.1 kWh	1 PEV 30 e-mi 10.2 kWh	1.5 PEVs 45 e-mi 15.3 kWh	2 PEVs 60 e-mi 20.4 kWh
$ -	$ 4,145	$ 8,289	$ 12,434	$ 16,579
$ 1,000	$ 2,707	$ 6,852	$ 10,997	$ 15,141
$ 2,000	$ 1,270	$ 5,415	$ 9,559	$ 13,704
$ 3,000	$ (167)	$ 3,977	$ 8,122	$ 12,267
$ 4,000	$ (1,605)	$ 2,540	$ 6,685	$ 10,829
$ 5,000	$ (3,042)	$ 1,103	$ 5,247	$ 9,392
$ 6,000	$ (4,480)	$ (335)	$ 3,810	$ 7,955
$ 7,000	$ (5,917)	$ (1,772)	$ 2,372	$ 6,517
$ 8,000	$ (7,354)	$ (3,210)	$ 935	$ 5,080
$ 9,000	$ (8,792)	$ (4,647)	$ (502)	$ 3,642
$ 10,000	$ (10,229)	$ (6,084)	$ (1,940)	$ 2,205

(b) $1.70/hour

Project cost	Utilization per day: 0.5 PEVs 15 e-mi 5.1 kWh	1 PEV 30 e-mi 10.2 kWh	1.5 PEVs 45 e-mi 15.3 kWh	2 PEVs 60 e-mi 20.4 kWh
$ -	$ 4,144	$ 8,287	$ 12,431	$ 16,575
$ 1,000	$ 2,706	$ 6,850	$ 10,994	$ 15,137
$ 2,000	$ 1,269	$ 5,413	$ 9,556	$ 13,700
$ 3,000	$ (168)	$ 3,975	$ 8,119	$ 12,263
$ 4,000	$ (1,606)	$ 2,538	$ 6,682	$ 10,825
$ 5,000	$ (3,043)	$ 1,101	$ 5,244	$ 9,388
$ 6,000	$ (4,481)	$ (337)	$ 3,807	$ 7,951
$ 7,000	$ (5,918)	$ (1,774)	$ 2,369	$ 6,513
$ 8,000	$ (7,355)	$ (3,212)	$ 932	$ 5,076
$ 9,000	$ (8,793)	$ (4,649)	$ (505)	$ 3,638
$ 10,000	$ (10,230)	$ (6,086)	$ (1,943)	$ 2,201

(c) $145/month/PEV subscribed

Project cost	Utilization per day: 1 PEV 15 e-mi 5.1kWh	1 PEV 30 e-mi 10.2 kWh	2 PEVs 45 e-mi 15.3 kWh	2 PEVs 60 e-mi 20.4kWh
$ -	$ 10,869	$ 8,302	$ 19,171	$ 16,604
$ 1,000	$ 9,432	$ 6,865	$ 17,734	$ 15,167
$ 2,000	$ 7,994	$ 5,427	$ 16,296	$ 13,730
$ 3,000	$ 6,557	$ 3,990	$ 14,859	$ 12,292
$ 4,000	$ 5,120	$ 2,553	$ 13,422	$ 10,855
$ 5,000	$ 3,682	$ 1,115	$ 11,984	$ 9,418
$ 6,000	$ 2,245	$ (322)	$ 10,547	$ 7,980
$ 7,000	$ 807	$ (1,759)	$ 9,110	$ 6,543
$ 8,000	$ (630)	$ (3,197)	$ 7,672	$ 5,105
$ 9,000	$ (2,067)	$ (4,634)	$ 6,235	$ 3,668
$ 10,000	$ (3,505)	$ (6,072)	$ 4,797	$ 2,231

achieved (at varying costs) through either temporal shuffling of cars in and out of the parking space or via multiplex shuffling of electrons flowing from the charger through multiple cords to multiple cars.

3.2 Fueling-Cost Benchmarks: MUD Charging and Gasoline Equivalents

This section provides benchmarks to facilitate the comparison of the driver costs of various fueling alternatives. Table 4 begins with the price of refueling at the four different per-hour recharging price levels described above for Table 1. Recall that pricing level 1 is a "breakeven" level that covers the cost of electricity. Table 4 then translates those prices into the dollars-per-electric-mile equivalent, using MUD-charging base-case assumptions. It also shows equivalent prices for electricity (per-kWh) and gasoline (per-gallon), representing the fueling alternatives facing a resident driver. Gasoline equivalents are shown both relative to a conventional vehicle (CV) baseline of 27.2 mpg and a sales-weighted plug-in-hybrid electric vehicle baseline of 41.1 mpg.

Table 5 adds the per-kWh and per-month pricing structures to the picture of the four different pricing levels described above. Recall that the cost-recovery potential achieved at each price level is nearly the same across pricing structures: "break-even" at pricing level 1, > $1,000 at pricing level 2, > $3,000 at pricing level 3, and > $5,000 at pricing level 4. As seen in the second column of Table 5, each price level is also roughly equivalent on a dollars-per-electric-mile basis—roughly 6¢, 10¢, 13¢, and 17¢ per electric mile, respectively.

Compared to a conventional vehicle, the "breakeven," $0.00/kWh-markup scenario in Table 5 illustrates that covering only the marginal cost of average residential electricity presents the resident driver with a low gasoline-equivalent price ($1.52/gal), and thus a large incentive to drive a PEV and charge at home. Even covering markups of up to roughly $0.20/kWh ("medium price") provides financial motivation, and a $0.30/kWh markup ("high price") is only slightly more than California gasoline. Recall that the $0.30/kWh markup was assumed as the

Table 4 Fueling cost benchmarks: per-hour MUD recharging

Pricing level[a]	$ per electric mile (/e-mi)	Electricity equivalent (/kWh)	Gasoline equiv., CV (/gal)[b]	Gasoline equiv., (plug-in) hybrid (/gal)[c]
1. $0.65/h actively charging	$0.06	$0.19	$1.72	$2.60
2. $1.00/h actively charging	$0.10	$0.29	$2.65	$4.00
3. $1.35/h actively charging	$0.13	$0.39	$3.58	$5.41
4. $1.70/h actively charging	$0.17	$0.49	$4.50	$6.81

[a] Each pricing level (1–4) provides the same amount of cost-recovery potential (Table 1)
[b] CV = conventional vehicle = 27.2 mpg [10]
[c] (plug-in) hybrid = 41.1 mpg [9]

Table 5 Fueling cost benchmarks: per-kWh, per-hour, and per-month MUD recharging

Pricing level[a]	$ per electric mile	Electricity equivalent (/kWh)	Gasoline equiv., CV (/gal)[b]	Gasoline equiv., (plug-in) hybrid (/gal)[c]
1. *Breakeven prices*			*"A Steal"*	*"Incentivizing"*
Electricity cost ($0.164/kWh, year 1)	$0.06	$0.16	$1.52	$2.30
$55/month	$0.06	$0.18	$1.71	$2.58
$0.65/h charging	$0.06	$0.19	$1.72	$2.60
2. *Low prices*			*"Incentivizing"*	*"Cheap"*
Electricity cost + $0.10/kWh	$0.09	$0.26	$2.45	$3.70
$85/month	$0.10	$0.29	$2.64	$3.99
$1.00/h charging	$0.10	$0.29	$2.65	$4.00
3. *Medium prices*			*"Cheap"*	*"Uncompetitive"*
Electricity cost + $0.20/kWh	$0.12	$0.36	$3.37	$5.10
$115/month	$0.13	$0.39	$3.57	$5.40
$1.35/h charging	$0.13	$0.39	$3.58	$5.41
Gasoline price (~CA 2012 average)	$0.15	$0.43	$4.00[d]	
4. *High prices*			*"Equivalent"*	*"Forget it"*
Electricity cost + $0.30/kWh	$0.16	$0.46	$4.30	$6.50
$1.70/h charging	$0.17	$0.49	$4.50	$6.81
$145/month	$0.17	$0.49	$4.51	$6.81

[a] Each pricing level (1–4) provides the same amount of cost-recovery potential (Table 1)
[b] CV = conventional vehicle = 27.2 mpg [10]
[c] (plug-in) hybrid = 41.1 mpg [9]
[d] http://articles.latimes.com/2013/jan/01/business/la-fi-gas-prices-20130101

base case as necessary for the MUD station owner/operator to cover roughly $5,000 of total project costs. Acknowledging that the driver may consider some additional private or social value (e.g., from the convenience of residential charging, parking or other associated benefits, increased zero-emission/oil-free travel, etc.), this nevertheless might represent a reasonable maximum markup that the MUD site host could expect residents to fully utilize in the near term.

Similar scenarios for each of the hourly and monthly price structures indicate that, compared to a conventional vehicle, an economic incentive exists for PEV drivers to use residential charging at prices at or below roughly $1.35/h or $115/month, respectively (scenario group #3, "medium prices").

An interesting picture develops in column 5 for the plug-in-hybrid driver, however, who has the option to forego recharging and use gasoline at any time. Because hybrids, including plug-in hybrids, are more efficient at utilizing gasoline, the costs of charging appear relatively less favorable to these drivers. Using the

Table 6 MUD recharging competitive price threshold, hybrid baseline[a]

1.1.1.1.1 Pricing level 2[b]	Gasoline equiv. (/gal)
Electricity cost + $0.10/kWh	$3.70
$85/month	$3.99
$1.00/h actively charging	$4.00

[a] (plug-in) hybrid = 41.1 mpg [9]
[b] provides $1,000–$2,000 worth of cost-recovery potential in all cases

sales-weighted, EPA-rating average of roughly 41.1 miles/gal, these drivers see each pricing group (1, 2, 3, or 4) as at least one group less attractive than it appears to the conventional-vehicle driver. For example, those drivers might only regularly fuel on electricity if prices are kept at or below price level 2—or $1/h, $85/month, or a $0.10-markup—as highlighted in Table 6.

The comparison is even worse for drivers that have a better-than-average vehicle at their disposal. For example, a driver of a plug-in Prius (50 miles/gal when on gasoline) would have a larger incentive not to use expensively-priced charging than a Volt driver, who in turn might view scenario-group-3 price levels as uncompetitive.

It should further be noted that several all-gasoline hybrids without plug-in capability (e.g., from Toyota and Ford) also achieve better gasoline efficiency than the plug-in-hybrid sales-weighted average of 41 miles/gal. To name the most popular example, the "regular" MY2012 Toyota Prius has an EPA-rated fuel economy of 50 miles/gal. Thus, even an all-battery EV driver might choose to drive their PEV less in favor of their regular gasoline-only Prius if charging prices are at scenario-group-3 levels.

3.2.1 Sensitivity and Uncertainty Analysis of Driver Cost Calculations

Following Sect. 3.1.1, sensitivity analysis was conducted on the inputs to the "electricity cost + $0.20/kWh" driver-fueling-cost calculation that produced $3.37/gal relative to a conventional vehicle. The additional key input not discussed in the previous uncertainty analysis is the conventional vehicle fuel economy, which was allowed to range from 22.8 miles/gal (described above) to 29 miles/gal (based on 3 years of historical change to allow for any increases that have started to occur after 2011 as consumer preferences change and new vehicle standards begin to take effect).

Monte Carlo and bounding analyses indicate the point estimates of fuel costs are reasonable, though the range produced extends to much higher (less competitive) gasoline-equivalent prices than illustrated in Table 4. Uncertainty in the costs of electricity contributed roughly 70 % of the uncertainty in the fuel cost estimate, whereas assumed electric vehicle efficiency contributed a little over 20 % and gasoline vehicle efficiency a little less than 10 %.

This analysis suggests that the benchmarks presented above are reasonable so long as electricity costs are near the average value of $0.1640. As electricity costs increase toward $0.37/kWh, the financial incentive to even the driver with an inefficient conventional vehicle vanishes.

4 Conclusions

This analysis finds significant opportunity for recharging facility cost recovery at prices that resident drivers might find financially motivating. Prices on the order of $0.36/kWh (including electricity costs and markup), $1.35/h-of-active-charging, or $115/month allow recovery of roughly $3,000–$4,000 in station investment per vehicle served under the baseline assumptions examined herein. This investment may not be sufficient to cover costs in a wide range of MUD environments (e.g., those that require parking-lot trenching or that have inadequate electrical panel capacity or long distances between the panel and desired charging locations). Further, these price levels may be considered uncompetitive to a sales-weighted average plug-in-hybrid driver. Plug-in-hybrid EV drivers with better-than-average vehicles—or even all-battery EV drivers with an efficient all-gasoline hybrid as a second vehicle—might be even less tolerant. Compared to a 50 miles/gal alternative, residential charging prices might have to be kept below $0.26/kWh, $1.00/h of active charging, or $85/month. These levels provide only roughly $1,000–$2,000 worth of cost-recovery potential per vehicle served.

For a given level of cost recovery, each pricing structure has unique characteristics. Per-kWh pricing benefits from the sales volume brought about by greater commute distances or other increases in utilization, per-hour pricing is negatively affected by higher charging power, and per-month profitability is subject to electricity-cost risk. Analysis of financial viability calculations indicates the per-kWh pricing structure offers significant upside potential while being less negatively affected by uncertainty in the inputs, both on average and in terms of minimizing variability in expected cost recovery.

Further differences are indicated here and could be explored in future work. The hourly rate structure has the disadvantage of potentially discriminating against older PEV models that charge more slowly and thus will effectively pay more per fill than will new PEVs. If not based only on the time spent actively charging, it may also discriminate against vehicles that do not require a lot of charge. For example, it may only take roughly 1.5 h to recharge a 15-mile electric range, even for older PEVs. Unless drivers move their cars or are not billed for the time after charging is completed, their costs per kilowatt-hour continue to rise, quickly reaching uncompetitive levels.

Both the hourly and markup fee structures come with the added costs of measuring and billing for the quantity of electricity or time that PEVs consume. As a flat-rate structure, the per-month method avoids these measurement and billing costs but has the disadvantages of both creating the perverse incentive to minimize charging and imposing different unit costs (e.g., cost per electric mile driven) on PEV drivers who travel differing numbers of e-miles daily.

Regardless of pricing structure, increasing facility utilization could significantly improve potential profitability. This is true across pricing structures, but is dramatic with each additional subscriber to the per-month structure. However, this might prove challenging given potential costs due to evening on-peak electricity costs and/or the need to shuffle cars. Low-cost solutions that increase utilization while minimizing per-vehicle installation and management costs (e.g., multiplexed, perhaps lower-power charging facilities) might help address these constraints, and should be a part of ongoing analysis to better understand the costs and benefits of implementing PEV recharging in MUDs.

Acknowledgments This work was supported in part by the South Coast Air Quality Management District (SCAQMD) and the generous general support of the UCLA Luskin Center for Innovation by Meyer and Renee Luskin. Precursor and related work was supported by the Southern California Association of Governments and the SCAQMD as part of Regional PEV Readiness Planning activities in California funded by the U.S. Department of Energy and California Energy Commission. The authors would like to thank these organizations and the individuals involved, as well Jon Overman, who conducted foundational analysis. However, the opinions, conclusions, and recommendations are solely those of the authors.

References

1. Lin Z, Greene DL (2011) Promoting the market for plug-in hybrid and battery electric vehicles: role of recharge availability. Transp Res Rec 2252:49–56
2. Lee T-K, Adornato B, Filipi ZS (2011) Synthesis of real-world driving cycles and their use for estimating PHEV energy consumption and charging opportunities: case study for midwest/U.S. IEEE Trans Veh Technol 60(9):4153–4163
3. San Román TG, Momber I, Abbad MR, Sánchez Miralles Á (2011) Regulatory framework and business models for charging plug-in electric vehicles: infrastructure, agents, and commercial relationships. Energy Policy 39(10):6360–6375. http://www.sciencedirect.com/science/article/pii/S0301421511005696
4. EURELECTRIC (2010) Market models for the roll-out of electric vehicle public charging infrastructure. Union of the Electricity Industry, Brussels
5. Schroeder A, Traber T (2012) The economics of fast charging infrastructure for electric vehicles. Energy Policy 43(0):136–144. http://www.sciencedirect.com/science/article/pii/S0301421511010470
6. Botsford CW (2012) The Economics of non-residential level 2 EVSE charging infrastructure. In The international battery, hybrid and fuel cell electric vehicle symposium (EVS26), vol 26. World Electric Vehicle Association, Los Angeles
7. EIA (2013) Average retail Price of electricity, quarterly. In: Electricity data browser, vol July 2013. 20 Sep 2013 ed. U.S. Energy Information Administration. http://www.eia.gov/electricity/data/browser/-/topic/7?agg=1,0&geo=000000000004&endsec=8&linechart=~~&columnchart=ELEC.PRICE.CA-RES.Q&map=ELEC.PRICE.CA-RES.Q&freq=Q&ctype=linechart&ltype=pin&pin=&rse=0&maptype=0
8. Ecotality EV Project Electric Vehicle Charging Infrastructure Summary Report, Sept 2011 to Sept 2012 (Ecotality 2012)

9. Williams B (2013) U.S. Plug-in electric vehicle (PEV) sales trends and analysis: Dec 2010–Aug 2013. UCLA Luskin Center for Innovation. http://innovation.luskin.ucla.edu/content/market-dynamics
10. EPA (2012) Light-duty automotive technology, carbon dioxide emissions, and fuel economy trends: 1975 Through 2011, EPA-420-R-12-001a. U.S. Environmental Protection Agency (EPA)
11. Williams B (2012) Second life for plug-in electric vehicle batteries: the effect of grid energy storage value on battery lease payments. Transp Res Rec 2287:37–43. http://trb.metapress.com/content/v006652807610n06/?genre=article&id=doi:10.3141/2287-08
12. Cready E, Lippert J, Pihl J, Weinstock I, Symons P, Jungst RG (2002) Technical and economic feasibility of applying used EV batteries in stationary applications: a study for the DOE energy storage systems program (SAND2002-4084). Sandia National Laboratories, Albuquerque, NM
13. DeShazo J, Ben-Yehuda A, Williams BD, Hsu V, Kwon P, Nguyen B, Overman J, Sarkisian T, Sin M, Turek A, Zarate C (2012) Southern California plug-in electric vehicle readiness plan. UCLA Luskin Center for Innovation, Los Angeles. (http://innovation.luskin.ucla.edu/content/market-dynamics)

Solutions and Business Models for Wireless Charging of Electric Vehicles

Axel Barkow, Gianni Campatelli, Riccardo Barbieri and Stefano Persi

Abstract This article describes possible scenarios for wireless charged electric vehicles, as they are discussed within the UNPLUGGED project. It gives a brief introduction into wireless technology and explains the three possible implementation steps for charging electric vehicles: stationary, static en-route and dynamic en-route. Based on these scenarios the impact of wireless charging on future mobility, especially in urban environments is discussed.

Keywords Wireless charging · EV mobility · Smart city

1 Introduction to Wireless Technology

The most common method for Electric Vehicle (EV) charging is the use of a cable plug-in system used either at home, at work or at specific charging stations spread across some cities. An alternative to this charging method is wireless inductive charging. With this charging method it is possible to just position your car on a charging point and start the contactless charging process without any additional effort.

A. Barkow (✉)
Forschungsgesellschaft Kraftfahrwesen Aachen mbH, Steinbachstr. 7, 52074 Aachen, Germany
e-mail: barkow@fka.de

G. Campatelli · R. Barbieri
Department of Industrial Engineering, University of Firenze, Via Santa Marta, 3, 50139 Florence, Italy
e-mail: gianni.campatelli@unifi.it

R. Barbieri
e-mail: riccardo.barbieri@unifi.it

S. Persi
Enide Solutions, St. Marina, 98, 2º-1ª, Barcelona, Spain
e-mail: stefano.persi@enide.eu

D. Beeton and G. Meyer (eds.), *Electric Vehicle Business Models*, Lecture Notes in Mobility, DOI 10.1007/978-3-319-12244-1_7

The UNPLUGGED project, co-funded by the 7th Framework Programme of the European Commission, aims to investigate how the use of inductive charging of Electric Vehicles in urban environments improves the convenience and sustainability of car-based mobility. In particular, it will investigate how smart inductive charging infrastructure can facilitate full EV integration in the urban road systems while improving customer acceptance and perceived practicality.

One promising possibility of wireless power transmission is inductive coupling. This idea is not particularly new. Nicola Tesla proposed theories of wireless power transmission in the late 1800s and early 1900s. Inductive coupling uses magnetic fields that are a natural part of a current's movement through a wire. Any time electrical current flows through a wire, it creates a magnetic field around the wire. Bending the wire into a coil amplifies the magnetic field. If a second coil of wire is placed in this magnetic field, the field can induce a current in the wire. Hence, charging an EV based on inductive coupling takes three basic steps:

1. Current from the wall outlet flows through a coil inside the charger on the infrastructure side, creating a magnetic field.
2. When EV is placed over the coil, the magnetic field induces a current in a second coil, which is part of the vehicle and is connected to the EV-battery.
3. This current recharges the battery (see Fig. 1).

This charging principle is quite new to EVs, but already well known is other appliances like electric toothbrushes or the Qi standard for wireless charging of mobile phones [1]. However, applying this technology to EVs raises new challenges like positioning and interoperability between systems.

The UNPLUGGED project agreed and is promoting from its early stage the following distinction between inductive charging technologies and related charging scenarios:

Stationary charging:

- The vehicle is not moving for a medium/long period of time (>5 min)
- The driver does not intend to use the vehicle in the short-term
- Scenarios: Parking at home/the office/the supermarket etc.

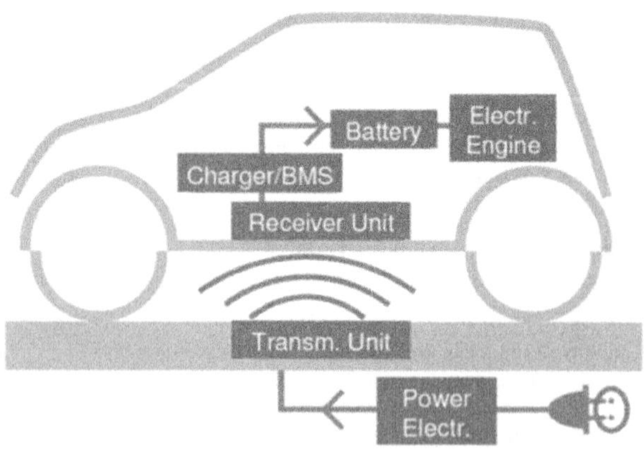

Fig. 1 Block diagram of the inductive power transfer system

Static en-route charging:

- The vehicle is not moving for a short amount of time (<5 min)
- The driver is likely still in the vehicle, on the way to his/her target location or intends to use the vehicle again very soon
- Scenarios: Traffic light, Bus stop, Taxi stand, Delivery truck

Dynamic en-route charging:

- The vehicle is moving
- The driver is on the way to his/her target location
- Scenarios: Highway, Strategically chosen roads

Currently car manufacturers and their suppliers development efforts are focused on stationary charging scenarios considering systems with only low power transfer capability leading to longer charging times. Despite several presented prototype systems and ongoing field tests, there are no mass-produced EVs equipped with wireless charging technology at this moment. However, on enquiry many OEMs confirm that next-generation vehicles will make use of this promising technology.

2 Wireless Vehicle Features

Comparing conductive to inductive charging, the wireless technology offers several advantages. It is easily operated and comfortable due to the automation of the charging process. Since there is no need to plug-in a cable the driver will not have to get out of the car and get his hands dirty especially in rainy, snowy or muddy environments, which is particularly important for commercial vehicles. Furthermore there is no risk that the user forgets to plug in the vehicle when parking, losing the capability of a fit range for the following use of the car.

The wireless system is much safer against vandalism, misuse/abuse and environmental influences (e.g. humidity), since all devices are encapsulated in the vehicle and the ground. In a world where copper has reached a significant price and criminals don't shrink away from making money out of railway's overhead contact lines, how sure can we be about vehicle's charging cables? These advantages already prove the usefulness of stationary inductive charging compared to the currently more common conductive charging infrastructure. However wireless charging will offer many more possibilities in the future. Taking a closer look at en-route charging, the future potential of this technology becomes clear.

Currently, technological limitations from performance issues are posing the biggest obstacles towards a major migration to electro mobility. The main focus here is on the battery. It is strange that this component, which basically provides the possibility of electric driving, is the main limiting factor at the same time. The total cost of an electric vehicle is mainly determined by the cost of the battery system. The bigger the capacity of the battery, the more the cost rises. On the other hand, the battery capacity is also limiting the driving range of the vehicle.

Unfortunately customers want to buy EVs with a high driving range for a minimum amount of money, which are obviously contradictory requirements. This is where wireless en-route charging comes in.

With inductive en-route charging, EVs could be charged while standing at the traffic light, the bus stop or the taxi stand. Within the short timeframe of the vehicle stop, it would be impossible for the driver to carry out the necessary handling for conductive charging, i.e. get out of the vehicle and plug in a cable, charge and pull the cable out as soon as the traffic starts moving again. With wireless inductive charging these short timeframes could be used to charge the EV and hence increase its range.

This charging method is called static inductive en-route charging, because the vehicle is standing still while charging. In addition, there is also the possibility to charge the vehicle while it is actually moving, which is called dynamic en-route charging. This charging method holds the potential of giving the driver virtually limitless range as long as he stays on paths specifically adapted for dynamic inductive en-route charging. This could lead to a reduction of battery size, which would both lower the weight and the cost for the vehicle. Of course on the other hand the infrastructure would be quite expensive and it has to be evaluated, if or when, dynamic inductive charging is needed and preferable to only static inductive charging.

Although wireless systems have many advantages for EV charging, the technology also poses potentially significant safety concerns such as electrical shock due to the high electrical power, high magnetic field exposure to the general and potential fire hazards [2]. These concerns are primarily due to the presence of large power levels, large electromagnetic fields and operation in potentially hazardous locations (for example, operation in garages with flammable materials). Safety and performance standards for wireless charging for EVs are currently under development. The automotive industry and other organizations are developing the technology and improving it not only from a performance perspective, but also from a safety standpoint. But how is it possible to prevent the electromagnetic field from potentially impacting human health? Engineers have spent many years perfecting the magnetic structure of the coils, which shapes a particular magnetic field very well and makes it very contained [3]. Using simulations as well as real life assessments the knowledge about magnetic field levels around the car increased a lot. Another area where a lot of simulation and testing is done is people with pacemakers or other electronic implants that would be affected by electromagnetic fields. A key point is the understanding of these medical devices and what sort of limits they are built to and shielded for. Since there is no set of rules that says wireless charging must meet certain regulations for pacemakers yet, this exchange of information is crucial.

While looking on these risks of the wireless charging technology, it is important not to forget that also conductive has no intrinsic safety. There one of the major challenges is the need to connect cords and sockets in often cramped conditions and in bad weather, which brings its own set of risks.

3 Toward a Wireless City: Vision of the Future

In our vision the wireless charging technology is the solution for the future urban mobility. Over the time many technologies have advanced in order to become wireless: the radio and the telephone are just some famous examples [4]. This technology is already becoming important for small appliances such as Smartphones and iPads. The application to vehicles will be a technical challenge but in the near future few people will have a "psychological barrier" to use a wireless charging vehicle. Mobility in future cities will be carried out using electrical vehicles, mainly public but also private, and it is not sustainable for the urban environment that all this fleet would be cabled to the grid [5]. Wireless technology provides a very high energetic efficiency but removes all of the constraints introduced by cables and, in the most advanced application—the dynamic charging—strongly reduce also the need of large batteries to obtain longer vehicle autonomy [6].

The provision of wireless charging in cities will progress in a number of steps. The first will be probably applications for private mobility, with charging devices that can be easily installed in private parking lots instead of traditional plug-in solutions. The second will be increased use of wireless technology in public spaces such as shared parking (for private mobility), bus and taxi stops, with the feasibility of traffic light charging eventually evaluated. The last step is the introduction of a dynamic charging system to provide power to vehicles in motion. This application will be a cornerstone of the mass introduction of EV based mobility but it will require necessary investments in infrastructure.

3.1 Static Wireless City

This first implementation step will be mainly for private mobility. The business model is the same as traditional EVs, but this solution could improve the diffusion of such vehicles improving the usability of the system thanks to the elimination of the need of a cable to charge the vehicle. This will provide many advantages such as: improved safety (no physical connection must be performed by the user), less invasive infrastructure (the charging device is small and could be integrated in concrete), no need to remember to plug in the vehicle at the end of its use, useful also for short stops [6, 7]. An example of the dimension of the charging devices is provided in Fig. 2, which shows two commercial products developed by Evatran [8] and WiTriCity [9].

It is important to note that these systems have a price that is only sensibly higher than the installation of a plug in system. Most of the cost is needed for the installation of the vehicle on-board pick up device. Regarding the system efficiency, this is really high if the vehicle pick-up device and the en route inductor are well aligned (a tolerance range of 10 cm would allow the maximum efficiency) [6, 10]. Within the UNPLUGGED project a study has been carried out to evaluate the

Fig. 2 *Left* Evatran's plugless power electric vehicle charging system [8] (Courtesy of Evatran corp.) *Right* WiTriCity system [9] (Courtesy of WiTricity Corporation)

Table 1 Parking accuracy without assisting system or using aftermarket solutions

	Test	Longitudinal distance		Transverse distance	
		Mean	Standard deviation	Mean	Standard deviation
Not aided stop	53	13.3	9.8	9.3	8.1
Visual system aided stop	39	11.0	8.6	11.6	7.9
Audio/visual system aided stop	39	7.0	4.5	7.9	4.7

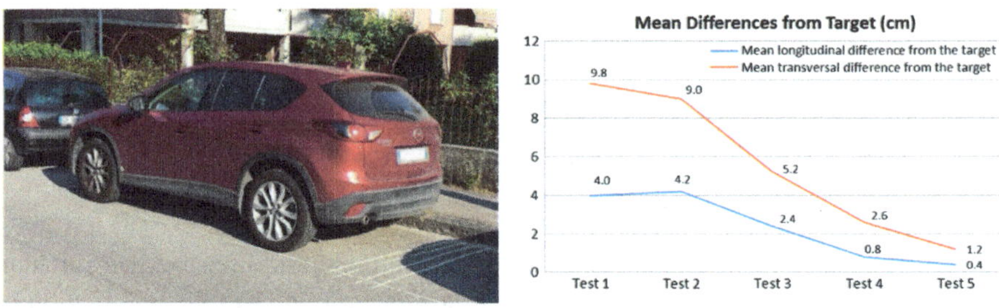

Fig. 3 Parking accuracy with an advanced driver assistance system

natural repeatability of car drivers in order to assess if this performance could be met by the nonprofessional road users. The tests have been carried out with a car without a parking assistant system and with a modern advanced driver assistance system (ADAS). The results have shown that without an assisting system a positioning error of about 14 cm could be obtained, as reported in Table 1, with the comparison of other aftermarket parking systems. The use of an already vehicle embedded ADAS, could make the repeatability of parking positioning reach a performance of 2–3 cm after short training of the driver, as reported in Fig. 3. This proves that for such technology the highest efficiency could be obtained with little effort by the driver and the technology is ready to support this scenario.

3.2 *Static En-route Wireless City*

Recharging in private areas or at public parking lots, however, is not the only possible application of the wireless power transfer technology. The next exploitation step for such technology is the introduction of recharge stations within the urban environment. These could be used both by public transportation vehicles (busses and cabs) and private cars. The main advantage of this technology would be the easier introduction of E-busses for urban mobility [11], some EU cities, such as London, are actually engaged in a test phase for this solution. This is an indication of how this technology is close to a large scale application. The recharge stations could be placed near the points where the bus stops for longer time (i.e. terminal stops) in order to recharge its batteries. This would solve the issue of service level, often limited by the bus autonomy, and would reduce the cost of batteries; a vehicle that could be charged at each terminal needs only a battery capacity to reach the following terminal and not, as usual, a battery able to satisfy the daily mileage. The reduction in battery size would also have beneficial effects on the vehicle weight and hence on the power consumed per kilometer. This approach could be easily used also for taxi cabs that often spend a relevant percentage of working time in dedicated waiting lanes. The possibility of fast charging their batteries within the route would allow for easier introduction of EVs, enabling drivers to achieve the needed daily mileage without requiring large and expensive battery packs. For example the autonomy of a Nissan LEAF that could be used like a taxi cab is 190 km with a standard driving cycle, a mileage often not enough for urban cabs [12]. In conclusion for public mobility this solution could be easily implemented but a detailed analysis of the location of the charging stations must be performed in order to have an economically sustainable solution. In the next section of this article a study is presented on the urban buses of the city of Firenze (Italy) which describes service level improvements and cost reductions.

For private mobility, this approach is not as easily applicable as is the case with public mobility. This is due to problems such as greater variability in vehicle positioning, shorter average stop times, misalignment at traffic lights, parking spaces constrained by the presence of other vehicles and traffic conditions in the street.

3.2.1 Case Study of Static En-route Charging for a Bus Service

Within the UNPLUGGED project, the bus service of the city centre of Firenze has been studied as a test case. The service, which already uses full electric vehicles, is operated by ATAF (*Azienda Trasporti dell'Area Fiorentina*), the bus society of Firenze, and is composed of three lines, C1, C2 and C3, with 5, 7 and 5 buses running respectively. This service is the only one within the city to use EVs because it services the historical area of the city that could not be accessed by traditional ICE vehicles. The historical city centre has many large pedestrian areas and its

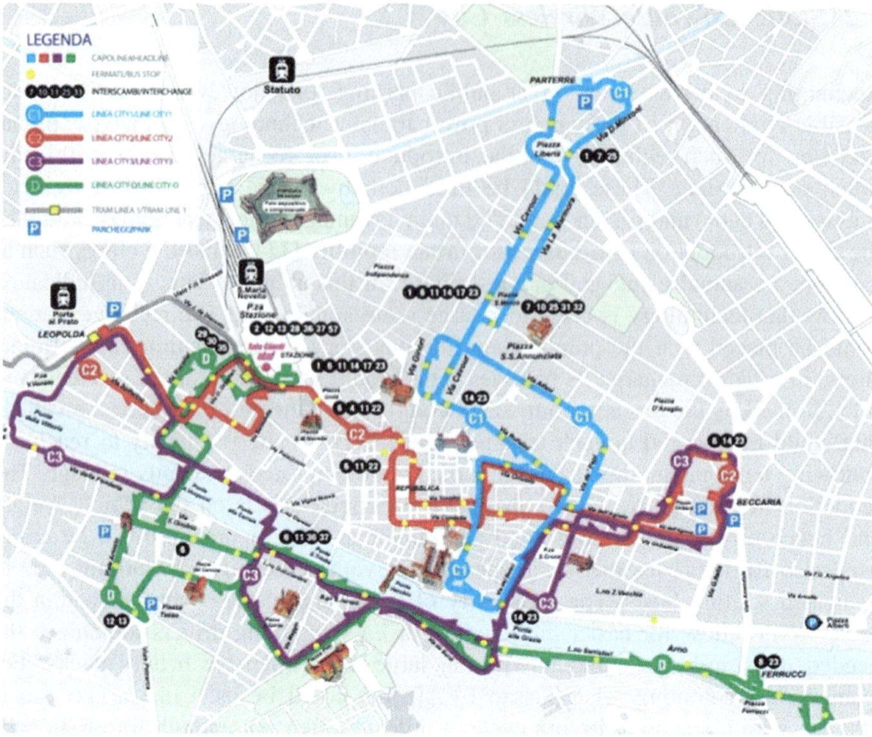

Fig. 4 Firenze city center bus lines

topography is due to its Renaissance origin with very narrow streets and many monuments and buildings of historical value, but also densely populated and visited by a very high number of tourists every day. For this reason the city council has been prepared to spend more than on traditional diesel buses; the price is however an ongoing issue and any solution that could reduce the upfront costs and those associated with the maintenance of batteries would be a key benefit and could provide a basis to extend this green service also to other areas of the city. Wireless battery recharging could provide an optimal solution to satisfy aesthetic, cost, mobility, air pollution and noise requirements.

Simulation has been used to determine the required number of recharging points and battery size for the buses. This models the behavior of the 17 buses running along the city (Fig. 4) checking the level of service provided and the total cost of the infrastructure/battery system.

The data needed by the model can be divided into two main categories, descriptive data of the functioning system, such as bus routes, spatial dispersion of the terminal stops, number of buses running on each line, and statistical data that describes the behavior of the system. This includes average consumption per trip, average waiting time at bus stops, average number of stops a bus will make on a

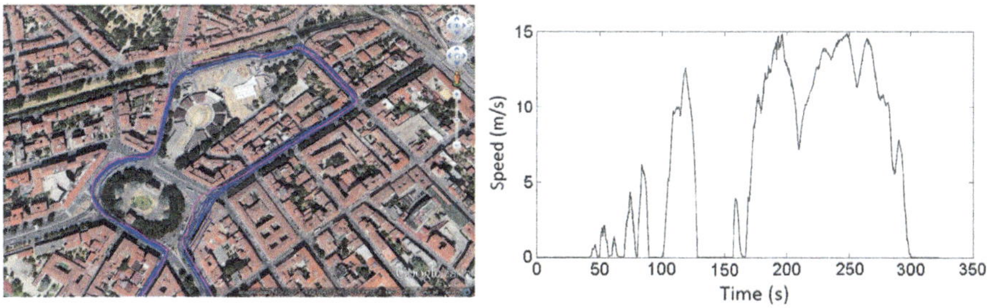

Fig. 5 Particular of a C1 path and relative drive cycle

defined route and average amount of time spent at the terminal stop. These datasets are collected from the analysis of drive cycles to establish variable distributions (Fig. 5).

The model has been developed on simulation software that uses the Monte Carlo method to determine pseudo-casual values for the variables [13]. The model developed is based on a balance equation of the SOC of each vehicle battery (1) that evaluates the battery behavior in each station for each vehicle:

$$SOC_{out} = SOC_{in} + PI_{capacity} * TIME_{stop} - DISCHARGE \quad (1)$$

where:

- SOC_{out}: the SOC at the end of each path between two consecutive stops.
- SOC_{in}: the state of charge (expressed in kWh) of the battery before being charged by infrastructure or discharged by driving cycle. This value is the result of the previous behavior of the system.
- $PI_{capacity}$: parameter that expresses the power inverter capacity at each stop.
- $TIME_{stop}$: variable that determines the stop time at a certain stop.
- DISCHARGE: variable that expresses a certain discharge value for a single path between two consecutive stops.

A cost function is then defined for a period of 20 years in order to optimally size battery and infrastructure:

$$\begin{aligned} Total\ Cost = & \sum_{j=1}^{n} (Battery\ cost_j + Battery\ Substitution\ Cost_j) \\ & + \sum_{i=1}^{m} Infrastructure\ Cost_i \end{aligned} \quad (2)$$

In Eq. (2) *j* represents the total capacity of batteries used in the control time period and *i* represents the recharge points developed.

The simulation was run for 365 working days, 13 h/day of service, with a power inverter capacity of 50 kW and 17 buses running on 3 routes. The first simulation

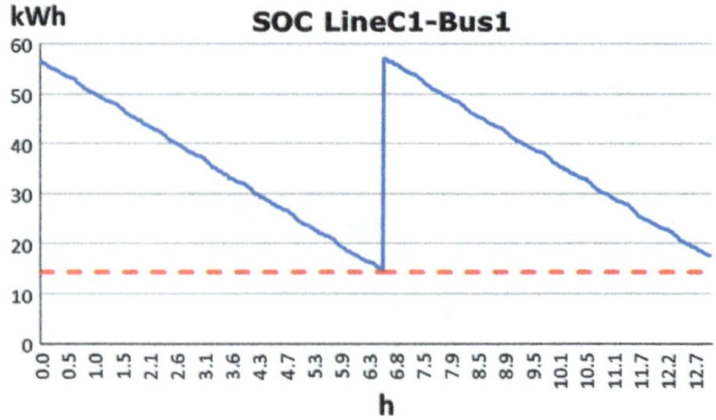

Fig. 6 Example of SOC trend for line C1 bus 1 in as-is scenario

was made on the *as is* situation, hence without any recharge system, but only with pure electric vehicles equipped with 71 kWh batteries. To protect the battery from deep charges or discharges, however, it has been considered that battery working range has to be 80–20 % and so 57–14.21 kWh as shown in the example of Fig. 6.

As in the real world, it is possible to see that at about halfway through the day, the battery falls under the threshold value of 20 % of total capacity and so the bus has to stop and must be replaced by another one with full battery. This issue, that happens once a day for all the routes and adversely affects both the service level provided to the customer and the cost effectiveness. As a result it is necessary to have a double fleet of electric buses to provide the service for a whole day.

The second simulation has been made with an on board battery of 3.75 kWh (protection limits 3.1–1.5 kWh) and a power inverter of 50 kW at each of terminal stop.

This scenario is the one with the least possible battery capacity that does not allow the SOC to fall down the lower protection limit during the simulated period (as shown in Fig. 7). The workload is all up to the infrastructure and the battery pack is only needed as a kind of energy buffer to reach the next recharge station.

The last scenario is opposite to the second one. The battery size is larger and the workload is distributed between battery and infrastructure; the bus starts the service

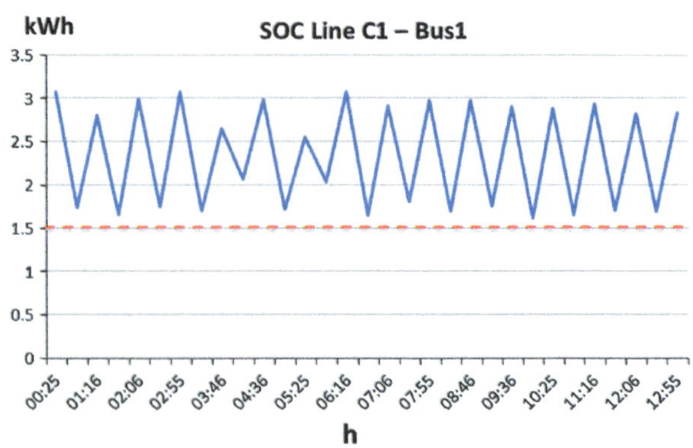

Fig. 7 Example of SOC trend for line C1 bus 1 in 3 kWh scenario

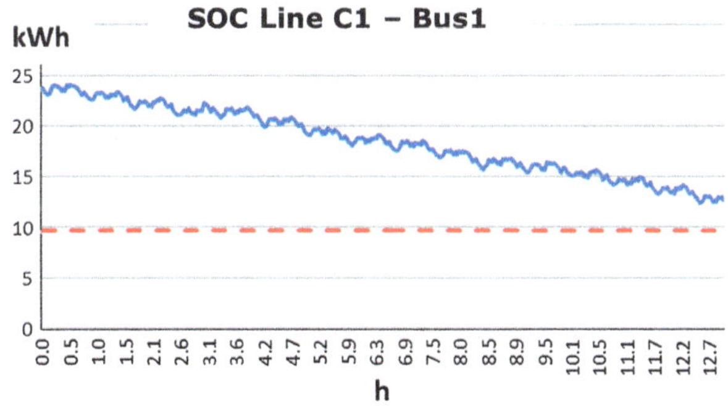

Fig. 8 Example of SOC trend for line C1 bus 1 in 24 kWh scenario

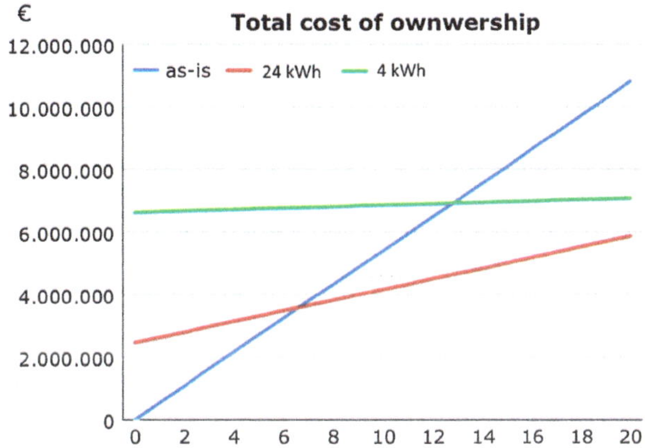

Fig. 9 Scenarios' total cost of ownership benchmark

in the morning with a fully charged battery and finishes with SOC very close to the lower limit, always without any stop during the daytime (see Fig. 8). Battery limits are 24–9.6 kWh.

To choose which of the proposed scenarios could be considered to provide the best total cost of ownership, each proposal has been assessed with the presented cost function (see Fig. 9). As previously stated, the amortization period considered has been set to 20 years.

Another study has been carried out for the whole bus system in the city. The total number of recharge stations has been evaluated and the effect on the electric grid has been estimated by ENEL, the Italian electricity provider. This study has shown that the implementation of this technology also at a city wide level is feasible: in fact, the power demand peak is less than 8 MW compared to the 9.000 MW of the rest of electric demand of the city. The only concern is that with this solution, most of the power consumption will be during daytime and not during the night as for private static recharge at home, as shown in Fig. 10. This will not allow the balancing of the power grid that usually has its peak power consumption during the day. Solutions to use a smart grid strategy to optimize the interface of larger EV fleet with the grid could minimize the problem.

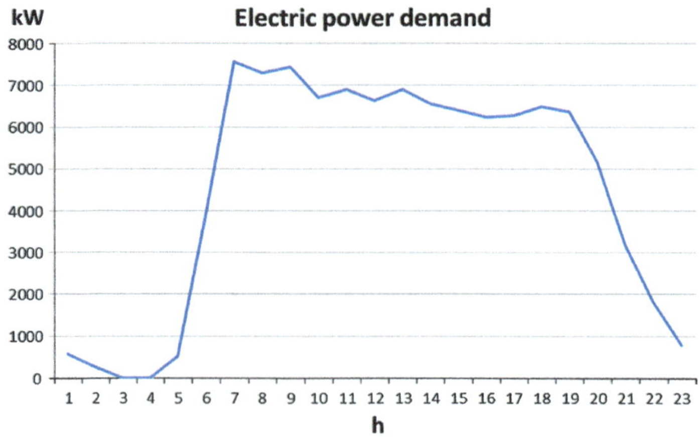

Fig. 10 Daily electric power demand for whole city wireless recharge infrastructure

3.2.2 Taxi Cabs

A similar analysis has been carried out also for the taxi cab service in the city of Firenze. A difficulty found is to have precise drive cycles for these vehicles due to the high variability of the routes. This makes it extremely difficult to use a simulated approach to determine the best infrastructure provision. So the global power need during a working day is estimated and the time for recharging the vehicle is assumed to be uniformly distributed during the day. The assumptions for recharging the vehicle and the system are reported in Table 2.

The total amount of recharge lanes and parking places within the city has been considered and geo-located.

Using the above presented assumptions, the number of E-cabs and the total amount of waiting lanes within the city of Firenze, it has been calculated that each E-cab has to wait at least 1.06 h/day in any recharge station in order to finish the working day without any vehicle failure for low battery level. This amount of time has been considered comparable with a cab daily routine by interviews with cab drivers of the city of Firenze. It is also possible to reduce the needed stop time or

Table 2 Features of vehicle and infrastructure for cabs simulation

Assumptions	
Nissan LEAF	Adopted vehicle
654	Number of cabs circulating within Firenze
35,000	Annual mileage for each vehicle
220	Working days per year
0.209	kWh/km: average Nissan Leaf consumption
3.7	Power inverter capacity
24	kWh capacity of Nissan Leaf battery pack
0.1	Electric vehicles percentage

reduce the power to be recharged at the end of the day, if multiple shifts per days are needed for the same vehicle, considering a higher power transfer of the inductor. Fast charge technology is able to recharge the on-board batteries using 22 kW power systems. This means that the battery pack of the vehicle could be fully charged in less than one hour and the needed fast charging within the route if a daily recharge at the base is provided would be less than 20 min. High power transfer is not an issue for inductive charging; the components must be properly aligned, but this solution is technologically feasible.

3.3 Dynamic Wireless City

A dynamic recharge system is able to charge the vehicle while it is moving [6]. This solution expands the concept of en-route charging to the whole travel of the vehicle. This is a really innovative technology that could virtually lead to an infinite vehicle autonomy (the vehicle is continuously powered by the grid) and reduce drastically the battery dimension (theoretically only a small buffer battery would be necessary). The main constraint to the introduction of such technology is the cost of the infrastructure and the low flexibility of the system. As soon as the infrastructure, constituted by a series of inductors or a litz cable, is placed under the road it is not possible to easily change the route of the vehicle. Within the UNPLUGGED project is studying how this system would affect the drivability of the vehicle and how much the battery could be reduced. Some technology advances are needed but these points are not critical for the feasibility of this approach. The most limiting issue is provided by the infrastructure that has a relevant cost, especially if the amortization must be carried out considering only few vehicles. With the mass introduction of electric vehicles this technology however would become more attractive, not only for the urban environment but also for longer distances such as highways, where the high number of vehicles would allow a fast amortization of the infrastructure.

The introduction of a continuous data exchange between the grid and the vehicles would also allow the introduction of many advanced services that could improve the usability of the vehicle and improve the safety, reliability and efficiency of the mobility system. Such technology would easily enable services like autodrive and car platooning: the vehicle will have a magnetic line to follow like the actual technology of AGV (Automated Guided Vehicles) already used in many industrial plants worldwide and a continuous exchange of information with the other cars; the introduction of a smart sensor on the vehicle and infrastructure would analyze the environment to verify the presence of people and other obstacles in its path. Although these are futuristic solutions, dynamic wireless charging could be actually applied efficiently in some areas, such as public mobility. The advantage of public mobility is that the travelled route is always the same during the day and there are many stops that increase the available recharge time. The authors have developed an analysis to evaluate the optimal percentage of the route to be electrified in order to obtain the optimal compromise between the cost of the

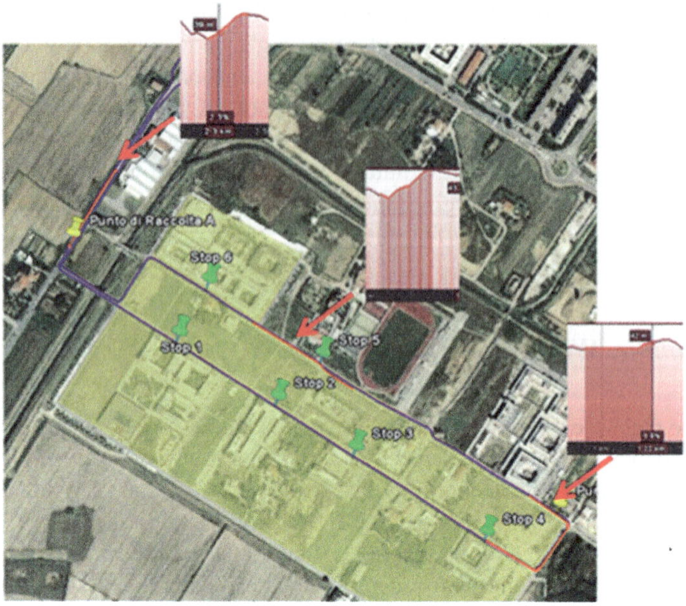

Fig. 11 Case study of bus for the Scientific Campus of UNIFI

infrastructure and the vehicle cost (with special regards to batteries). This percentage is around 30 % and is dependent on the route elevation and planned stop time at the terminals. The guidelines to define which would be the part of the route to be electrified are to use the part where the vehicle has a lower speed, that means longer time to recharge, and where the power need is greater, such as when climbing a slope. An example of the optimal electrified route is provided for a bus that services the Scientific Campus of the University of Firenze, Fig. 11. The algorithm to define the optimal length of route to be electrified takes into account the cost of the infrastructure, the cost of the batteries to be installed and the effect that the batteries duty cycle will have on their life. In general the system avoids deep discharge of the batteries because it has a negative effect on their life and on their maintenance cost. The solutions identified need 286 m of electrified path, with a 100 kW power inverter, and a bus with 7 kWh battery capacity.

4 Available Service for Users

The introduction of a wireless charging technology within an urban environment would allow the development of advanced services for the road users. The advantages of a continuous Vehicle-to-Grid (V2G) connection has been presented in Sect. 3.3 but also a simpler implementation of this technology, such as the static en-route charging scenario, would allow some special services. First of all the grid

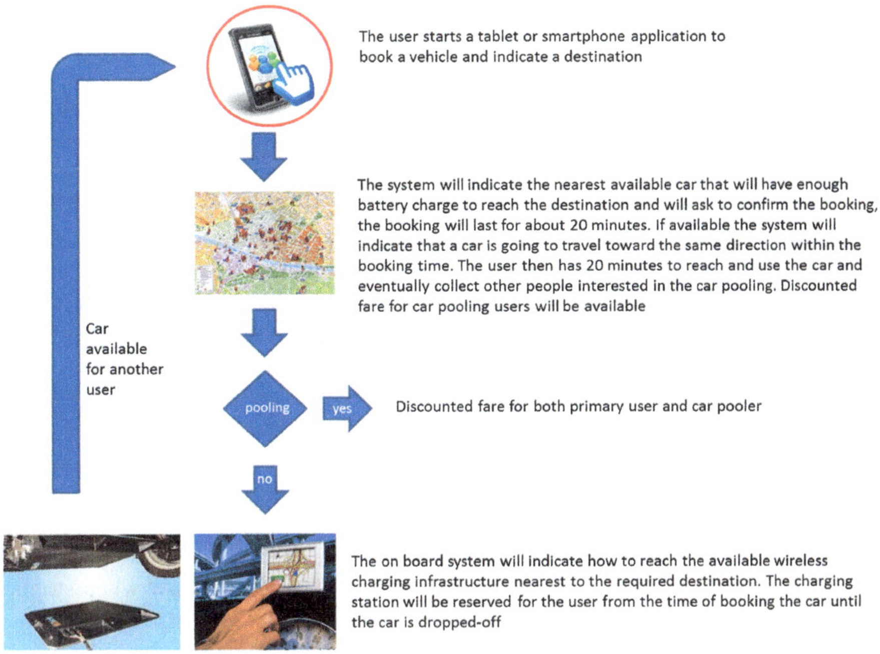

Fig. 12 Flow diagram of car sharing and pooling supported by a distributed wireless recharging infrastructure

of recharge stations could be managed like a large smart grid. Each station could be monitored and controlled by a central system and could interact with the users thanks to a city wide mobility server. From the grid point of view the vehicles could be used to level the instantaneous power request using their batteries as power damper. From the user's point of view, a V2G communication system could allow an easier introduction of advanced services such as the car sharing and car pooling. A vehicle connected to the grid is constantly monitored and every user could verify the vehicle autonomy or book a charging station near its destination choosing from the ones available in the distributed wireless charging infrastructure of the city. If the system knows a final destination it will become possible to search for possible car poolers in order to share the car. This solution could be supported by economic incentives and could be managed as reported in Fig. 12.

5 Conclusions

In conclusion, it is possible to say that the electro-magnetic coupling for charging EV batteries is an interesting possibility to increase adoption of electric mobility in cities thanks to the intrinsic safety and usability characteristics of the system itself.

The impact on human health of high power magnetic fields is still a quite open issue, however many studies are underway in the scientific community to understand how to eliminate, or at least limit the impact on human health.

Regarding the introduction of the technology to an urban environment, this article has reported the sustainability of wireless charging respect to infrastructure cost, environmental impact and ease of use. Moreover the implementation of such technology should be progressive, responding to the increased adoption of EVs, and potentially introduced into three stages: static, static en-route, and dynamic electric charging. The first scenario describes the introduction of recharge bays only in private areas, such as garages or public parking lots, where the car will be parked for more than 5 min, and proves how the existing park assistant systems would be enough to ensure the correct position to have good recharge efficiency. The second scenario describes the introduction of a bus and taxi cab service, within the case study city of Firenze, with a "recharge at the stops" technology. Thanks to the particular service model of public mobility, that envisages quite long stops in well-defined areas, the introduction of this technology could allow a reduction in battery size and lower the general cost of ownership whilst maintaining the same service levels. The last scenario, the dynamic wireless city, describes the future development of magnetic coupling, where vehicles will be charged not only when stopped but also while driving. This full integration within the city will also provide other advanced services for road users, such as V2G and car pooling/car sharing services, making the city a *smart city*.

References

1. http://www.wirelesspowerconsortium.com/
2. Jiang H, Brazis P, Tabaddor M, Bablo J (2012) Safety considerations of wireless charger for electric vehicles—a review paper
3. Ahn S, Kim J (2011) Magnetic field design for high efficient and low EMF wireless power transfer in on-line electric vehicle. In: Proceedings of the 5th european conference on antennas and propagation (EUCAP), 2011
4. Cannon BJ, Hoburg JF, Stancil DD, Goldstein SC (2009) Magnetic resonant coupling as a potential means for wireless power transfer to multiple small receiver. IEEE Trans Power Electron 24(7):1819–1825
5. Yamagata Y, Seya H (2013) Simulating a future smart city: an integrated land use—energy model. Appl Energy 122:1466–1474
6. Suh I-S (2011) Application of shaped magnetic field in resonance (SMFIR) technology to future urban transportation. In: CIRP design conference, 2011
7. Suh N-P (2010) Design of on-line electric vehicle (OLEV). In: CIRP design conference 2010
8. http://www.pluglesspower.com/
9. http://www.witricity.com/
10. Sallan J, Villa J, Lombart A, Sanz J (2009) Optimal design of ICPT system applied to electric vehicle battery charge. IEEE Trans Ind Electron 56(6):2140–2149
11. Steinhilber S, Wells P, Thankappan S (2013) Socio-technical inertia: understanding the barriers to electric vehicles. Energy Policy 60:531–539

12. Kempton NPW, Guensler R, Elango V (2011) Electric vehicles: how much range is required for a day's driving? Transp Res Part C: Emerg Technol 19(6):1171–1184
13. Metz M, Doetsch C (2012) Electric vehicles as flexible loads—a simulation approach using empirical mobility data. Energy 48(1):369–374

Part III
Energy Systems

Electric Vehicles as Grid Support

Kristian Handberg and Gill Owen

Abstract The vision of Electric Vehicles (EVs) providing energy back into the grid is one that has captivated people since the re-emergence of EVs and the rise of the Smart Grid. In this vision, EV owners, electricity network operators and society more generally share the benefits of smoothing the peaks and troughs of electricity demand and improving EV battery utilization. Whilst technical obstacles remain, the greatest limitations are those presented by electricity market rules. Within existing frameworks, charging can be managed according to network needs for widespread benefit. And more immediately, vehicles can supply emergency back-up power or act as local storage for renewable energy through interactions that take place behind the electricity meter. Successful deployment of these scenarios will serve as a stepping stone towards the future vision of EVs as grid storage.

Keywords Electric vehicles · Electricity networks · Charging · Vehicle-to-grid · Demand response

1 Introduction

The resurgence of electric vehicles has coincided with a period of unparalleled change in the electricity sector. Advanced monitoring, analysis, communications and control, distributed energy resources and various other innovations are being adopted at the local, regional and national level for efficiency improvements, improved energy access and reliability, and reduced environmental impacts.

K. Handberg (✉)
Percepscion Pty Ltd, Level 10, 99 Queen St, Melbourne, Australia
e-mail: khandberg@percepscion.com

G. Owen
Monash University, Building 74, Wellington Rd, Clayton, Australia
e-mail: gill.owen@monash.edu

D. Beeton and G. Meyer (eds.), *Electric Vehicle Business Models*,
Lecture Notes in Mobility, DOI 10.1007/978-3-319-12244-1_8

The coincidence of these technological developments have given rise to the vision of a fully-integrated electricity and transport system, where electric vehicles provide convenient storage of cheap, clean electricity for later re-use. This vision addresses the previously inefficient allocation of resources that is stationary vehicles, while at the same time facilitating greater integration of intermittently-generated renewable energy.

This compelling ideal has captured the public imagination beyond simply futurists, academics and technophiles. The endgame for some—to go “off the grid” and be free from the yoke of utilities and oil companies—is encapsulated by the prospect of an EV parked in the driveway of one’s home under a roof covered in solar cells.

However, to realize this vision a range of technical, economic and regulatory obstacles must be surmounted. Integrating vehicle batteries into grid operation; developing business models that address capital-intensive technology investments; and deploying solutions within current regulatory frameworks are significant challenges that will not be addressed overnight.

Progress is being made. Lessons learned from the first phase of EV deployments are informing technical design and standards for the next-generation of grid-interactive vehicles. Niche applications that could potentially support a positive business case for vehicles as storage are being explored. Electricity network operators are gaining confidence in their ability to accommodate, and even benefit from, electric vehicles.

This article steps through the features of electricity grids and electric vehicles relevant to their interaction. It explains the likely use-cases for managed vehicle/grid interactions based upon the overlapping characteristics of the two technologies and their operational environments. And through an improved understanding of the issues and opportunities, the article will set out pathway towards realization of the integrated-system vision set out above.

2 How Do Electricity Grids Work?

In many people’s mind “the grid” is a proxy for the system—electrical or otherwise—encompassing the society in which they live. However for the purposes of this article, the concept of a grid is defined in terms of the building, precinct, or regional-scale network connecting electricity sources with end-uses.

Design and operation of electricity networks revolves around the (often competing) objectives of reliability, affordability and sustainability.

Reliability reflects the continuous provision of electrical supply of the appropriate quality. When demand exceeds supply, system failures in the form of outages occur. When the quality of electrical supply goes outside of agreed limits (e.g. for frequency or voltage), equipment malfunctions (e.g. machinery running slower) or even failures can result.

Efforts to ensure system reliability and sustainability meet required levels must be considered alongside the costs associated with these investments. To ensure affordability, community-scale electricity is generally supplied within a market framework that seeks to balance costs and service levels, now and in future. Commonly this will place decision-making for network prices with an independent regulator, who will consider investment proposals from monopoly network operators on behalf of customers. As electricity networks are capital-intensive assets operated over decades, extensive regulatory frameworks provide clarity for all parties around the rules, costs and benefits for their interaction.

Activities such as on-site (distributed) energy generation are regulated in terms of their interaction with the overarching network. However, where activities are entirely "behind-the-meter" (refer to Figs. 1 and 2), decision-making is largely in the hands of the customer. By taking ownership of the reliability, affordability and sustainability of their network, the customer can strike a different balance between these objectives from the more utilitarian approach taken on their behalf in the regulated market.

An important influence on the design of electricity networks is the forecast needs of users. Accordingly, peak demand is a major driver for network investment [1]. Demand management is increasingly seen as an important tool to reduce the need for network investment and minimize the costs of electricity for users. Regional-scale networks pursue demand management programmes to varying degrees, primarily as a reflection of market rules which incentivize or obligate them to do so.

The majority of the benefits from demand management are obtained by focusing on managing loads during relatively infrequent periods of peak demand [2]. And as part of efforts to minimize transaction costs, demand management has traditionally focused on a relatively small number of large loads—for example commercial/industrial energy users rather than residential.

The digitalization of electricity networks is providing enhanced visibility, decision-making and control as part of the move towards "Smart Grids". In parallel, pricing structures are being implemented that better incentivize demand management. Technology cost reductions for renewable energy generation and energy storage are driving uptake and market reform. These developments are in pursuit of better outcomes relative to the system objectives of reliability, affordability and sustainability.

3 What Are the Characteristics of Electric Vehicles Relevant to Grid Support?

For the purposes of this article, an electric vehicle is defined as a vehicle that draws some or all of its propulsion energy from an external source via a plug (i.e. this definition excludes conventional hybrids). Although the latest incarnation of EVs as mass transit vehicles is relatively new and fast-evolving, they have some defining characteristics that are relevant to their interaction with electricity networks.

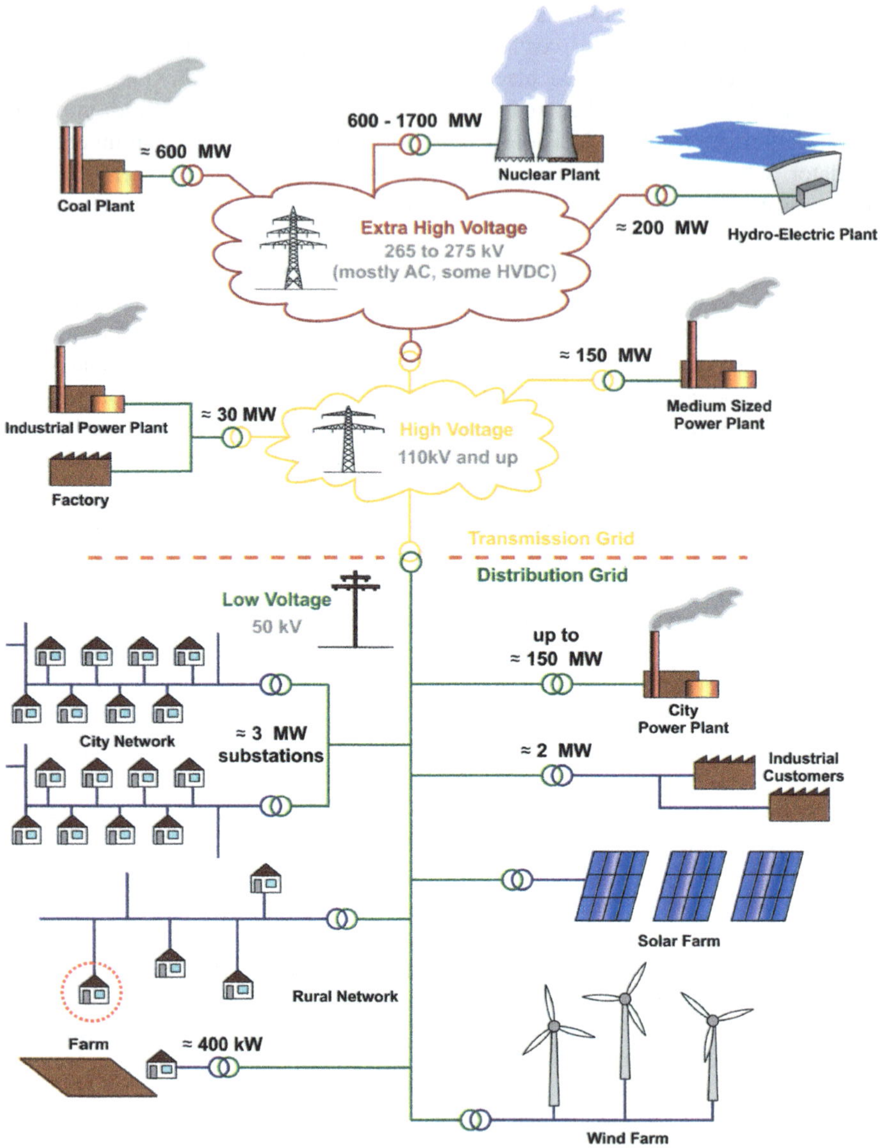

Fig. 1 A regional-scale electricity network of interconnected sources and end-uses that corresponds to the most common definition of "the grid", where a single home has been circled to provide context for Fig. 2

A key issue for vehicle/grid interactions is the primary application of the vehicle as transport. When and how far a vehicle is driven or planned to be driven determines when and how long the vehicle is plugged in, how quickly it is expected to charge, and how much charge it draws or is available to supply. Generalizations

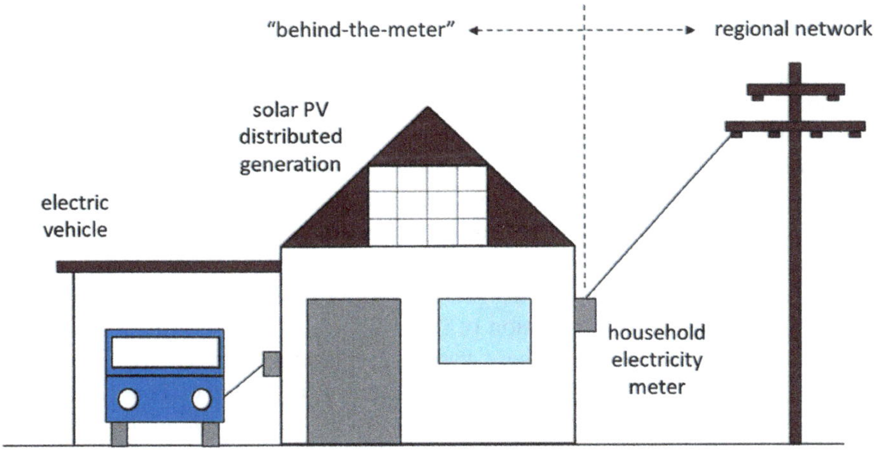

Fig. 2 A residential electricity network including grid-connected distributed generation and a plug-in vehicle. This network exists "behind-the-meter"—the demarcation point with the regional network as illustrated in Fig. 1

based upon the experiences and opinions of the first generation of EV owners include [3, 4]:

- Private vehicle owners primarily charge overnight by plugging in upon their arrival home and unplugging prior to setting off the following morning.
- Charging behaviour by private vehicle owners is responsive to price signals such as off-peak tariffs or free charging opportunities through the workplace, but is ultimately subservient to the vehicle utility.
- Fleet vehicles are highly variable in terms of their distance travelled and charging patterns, and there is a disconnect between fleet vehicle users and those with responsibility for electricity costs within fleet operations.
- Distances travelled between charging events are significantly less than the potential range of the vehicle when fully charged, and charging events are completed within the time available before the vehicle is next required for use.
- Quick-charging provided as a commercial service is used for emergencies rather than as the primary charging strategy.

While these generalizations are useful, it is important to note they are not strict rules as behaviour can vary markedly as a reflection of vehicle use requirements.

The storage capacity of electric vehicles varies—from 16.5 kWh in a GM Volt plug-in hybrid EV [5], to 24 kWh in a Nissan LEAF fully-electric vehicle [6], 85 kWh for the largest battery size offered in a Tesla Model S [7], and 324 kWh for a BYD electric bus [8]. Given that the average American home uses around 30 kWh of electricity per day [9], a plug-in passenger vehicle has the potential to add substantially to household electricity demand, or conversely to provide energy for several days if rationed.

The traction battery is the single most expensive component within an electric vehicle, costing thousands of U.S. dollars to replace [10]. While under normal operating conditions the battery will last for around 8–10 years [11], there is the potential for battery life to be reduced by variables within the operating environment. This may translate to warranty restrictions that re-allocate risk for owner-determined operating decisions, accelerated vehicle depreciation or a general reluctance from manufacturers to embrace vehicle-as-storage applications.

Battery packs prefer to operate within a relatively narrow temperature range, and can be damaged irreparably by extremes of temperature. In addition to the influence of ambient temperature, the electrochemical reaction that is at the core of battery technology results in the production of heat.

The complex interaction of these variables is managed by the vehicle Battery Management System (BMS), and translates to varying rates of charge/discharge within a single cycle and across separate cycles. Even within the operating parameters permitted by the vehicle BMS, battery life will be decreased by regular deep-discharging and/or fast-recharging—potential examples of owner-determined operating decisions.

At the high end of the variation within the charging cycle, peak power levels for EV charging can be significant. Home charging peak power loads are generally one of either 3.3 or 6.6 kW [6] but may go up to 20 kW [7]—any of which will be a significant addition to household electricity demand. This is significant due to the correlation between charging behaviour and household energy demand in the absence of other influences.

Commercial quick-charging (12–50 kW) [12], or charging for electric vans, trucks or buses (up to 200 kW) [8] is more significant again in terms of its potential impact upon the grid.

Charge management technologies are also relevant to consideration of EVs for grid support applications. Interactions between the vehicle, network and user systems will rely upon consistent application of technical standards at each interface. Vehicles may be equipped with (proprietary) charge management technology, including charge timers and smartphone applications that provide remote charge monitoring and management. Where absent from the car, dedicated charging equipment or building energy management systems may provide this capability.

For use as storage, vehicles must be capable of exporting energy back into the grid. Furthermore, the charge supplied must be compatible with the grid, and managed so as to reflect the needs of both the grid and the vehicle owner, within the constraints of the vehicle BMS and battery life impacts.

Bi-directional vehicle charging and charging infrastructure systems are emerging from technology providers [13, 14]. These systems deal with many of the technical challenges of extracting charge and supplying it into the grid, and are being tested within real-world operating environments. Additional work is underway on standards to assist with the regulatory issues associated with supplying energy into the grid [15] and integration with network management systems [16].

4 How Could EVs Provide Grid Support?

In considering the characteristics of plug-in vehicles as they relate to grid support, the scope of this article has been refined to focus specifically on intact and operational vehicles. This excludes applications where the traction battery is removed from the vehicle and utilized as stationary storage under "battery swap" operating models or as part of a "second-life" beyond its first life in the vehicle.

Building on this, the potential grid support applications can be considered in the context of a range of high-level costs and benefits.

The costs include the impact upon the vehicle utility, the impact upon battery life/vehicle re-sale value, and the technology (capital) and transaction (operating) costs for all stakeholders.

The benefits relate to avoided costs associated with grid investments to deal with peak demand increases, improved network capacity utilization from charging demand management, and potential uses and benefits arising from electricity stored in and sourced from the vehicles.

Once applications have been identified where there is an overall net benefit, the distribution of costs and benefits must be managed equitably within the system. For ease of implementation, this should happen through existing regulatory and market arrangements. Key enablers may include market rules which provide incentives for demand management or allow for stored energy to be supplied into the market. In the absence of enablers for market participation, behind-the-meter applications may be considered as a closed system.

Within this conceptual framework a number of applications have emerged that are presented in order of their ease of deployment.

4.1 Off-peak Charging

As outlined in Sect. 3, electric vehicle charging has the potential to add significantly to peak demand at the household level and in aggregate for regional-scale networks. If unmanaged, studies have found that network augmentation may be required once EVs are adopted by 10 % of households [17].

EV charging behaviour has been found to be responsive to price signals [18]. Time-of-Use (TOU) rate structures incentivize electricity to be used in off-peak periods—a form of demand management known as Demand Response [19].

EV charging activities are responsive to TOU incentives on account of the low impact on the driver in terms of their vehicle use, the technology solutions which enable easy and convenient charge management, and the significant benefit to them in terms of reduced electricity costs. Drivers respond to TOU rates by programming their charging to begin at the onset of the off-peak period—minimizing their costs without impacting their vehicle utility.

For the network operator, off-peak charging can be achieved through the provision of a TOU tariff as part of a DR programme. Historically, DR programmes have focused on commercial/industrial customers, however in many markets programmes now also apply to residential customers. The benefits relate to avoided investment in generation and/or network infrastructure to support additions to peak demand arising from EV charging. These savings translate to lower costs for all electricity users.

4.2 Optimized Charging

Optimized charging activities using Load Control (LC)—another form of demand management—allow the peaks and valleys of electricity demand to be addressed and the impacts of intermittent supply from renewable sources to be mitigated (refer also to Sect. 4.4 below). By more closely matching demand with supply and/or reducing demand fluctuations, supply interruptions may be managed, network asset lifetimes can be prolonged and maintenance costs reduced.

In addition to market rules that allow for and incentivize LC interactions, end-to-end technology solutions are required for charging to be managed from the network perspective. Vehicle charging loads should be visible and manageable to the network operator, who can then optimize charging in line with the wider network needs. This is most easily achieved via grid-integrated charging infrastructure that forms an extension of the Smart Grid.

The costs for an end-to-end technology solution have been found to be around one tenth the cost of network augmentation under an unmanaged charging scenario [17]. By optimizing vehicle charging, generation can be shifted to cheaper plants and influences the choice of new plants [20]. Studies have shown that by optimizing charging, over 50 % of households might adopt EVs without network augmentation being required [17].

However, a threshold issue for optimized charging in regional-scale grids relates to the ability and incentives for network operators to pursue demand management approaches. In addition, whoever makes the upfront investment in the grid-integrated charging solution needs to be able to recover the costs as part of the benefits distribution. For example, Italian market rules define grid-integrated charging outlets as part of the cost-recoverable utility asset base of the network operator [21].

Results from field trials suggest that driver acceptance of network-managed charging reflects such issues as their ability to override the managed charging activity (even at a cost), remote visibility and control over their charging, and information and clarity in relation to the managed charging activities (for example, that their vehicle will be fully charged by no later than 7 am) [22].

With the advent of advanced network control systems, transaction costs may now be reduced to allow the wider benefits associated with optimized charging to be realized [19]. In addition, the negative impacts associated with the creation of a "second peak" in electricity demand coinciding with the beginning of the off-peak tariff period may be avoided.

Within the LC framework drivers will manage their vehicle charging and use to avoid cost penalties and maximize benefits, but may elect to override LC events should their situation require it. For LC programmes focused on load reduction from air-conditioners or heating units during peak demand periods, the consumer typically receives a benefit of $USD 50–100 / year as a credit on their electricity bill [2]. This may be reduced or even exceeded by penalties associated with opting out of the LC events.

Optimized charging may facilitate renewable energy integration through closer matching of demand with supply [23]. For site owners with distributed generation, the EV can provide an alternative to feeding surplus electricity back into the grid [24]. This may be particularly relevant in the absence of a feed-in tariff for excess electricity [25].

4.3 Grid Storage for Emergency Back-up

As outlined in Sect. 3, electric vehicles have sufficient storage capacity to provide emergency power for several days if rationed in the event of wider system failure. This is a potentially high-value, even life-saving, application of electric vehicles as storage. In addition, as the energy is used local to the vehicle, transaction costs are minimal. This opens up the possibility of using EVs as a standalone "off-grid" power source, for instance during recreational camping.

In its simplest form, the use of EVs as a source of emergency back-up power can be achieved through devices that allow power to be supplied direct from a vehicle to small appliances, such as lighting or mobile phone chargers. This equipment generally converts the DC charge from a vehicle into a useable, if small, AC power supply approximating mains electricity independent of a hardwired electricity network [26, 27].

Once a vehicle is connected to a network, the appropriate controls, switches and protections must be present for it to act as a source of supply [28]. The network being served must be isolated, and the use of emergency power must be easily and clearly made known to avoid endangering anyone at work on the surrounding network, for example line-workers who are attempting to restore mains power [29]. Advice is generally available from the regional network operator on applicable standards and regulations.

At the top end of the scale in terms of cost, complexity and capability are standalone network operations that incorporate distributed energy resources (DERs). Microgrids—independently capable and controllable networks which include DERs and potentially storage—are of growing interest as a means of providing system resilience [30]. The inclusion of EVs as storage within these networks may ultimately prove to be a more important application than for the macro-grid [31].

4.4 General Grid Storage

Utilizing stationary electric vehicles as a storage facility for the grid is a simple, yet powerful vision. This model is what many consider to be the definition of Vehicle-to-Grid (V2G), a term which may apply to regional-scale network interactions alone, or may also include smaller system interactions known as Vehicle-to-Building (V2B) or Vehicle-to-Home (V2H). The challenge facing proponents of this vision lay in the investment risk arising from the uncertain benefits, particularly when considered alongside non-vehicle storage solutions.

The technical solution to utilize electric vehicles as grid storage is an enhanced version of the network-connected emergency back-up solution from Sect. 4.3. Regular, bi-directional energy flows can be managed using control systems that are evolving for more widespread application. Complex, automated monitoring and decision-making is increasingly possible through Smart Grid [32, 33], Building Energy Management (BEM) [34] and Home Energy Management (HEM) [35] systems. As outlined in Sect. 3, charge extraction on the vehicle-side will occur within the constraints of the BMS and through consistent application of technical standards each side of any sub-system interface.

The complication within this system relates to the control strategy that must reconcile the constantly-changing and potentially-competing uses of the vehicle as transport and grid storage. The response to this dynamic is constrained by the time required for meaningful charge transfer to take place—itself a variable controlled by the BMS.

As outlined in Sect. 4.2, drivers may be comfortable with managed charging of their vehicle so long as they retain ultimate control to ensure they have charge when they need it [22]. Furthermore, research suggests people discount potential revenues from V2G contracts heavily, and associate V2G with high inconvenience costs [36]. While these findings may be unsurprising in a market with practically no knowledge of V2G arrangements or operation, they highlight the challenges inherent to the V2G operating model.

The design of demand management programmes may provide some insights into the likely path forwards for V2G. At the outset, market rules must enable and incentivize the participation of energy storage, including from vehicles [37]. Under this scenario, grid storage capacity must be committed ahead of when it is provided. This will necessitate high confidence levels around the amount of charge that may be supplied in line with scheduling by the grid operator—a requirement that may conflict with the primary use of a vehicle as transport. Service agreements designed around the likely overlap between grid needs and vehicle charging connection may help address this and improve the potential for EVs to provide meaningful grid storage capacity.

Demand management programmes focus on interactions that minimize transaction costs relative to the value of the energy exchanged. For V2G this may mean corporate vehicle fleets able to be treated as a single storage facility, commercial vehicles with relatively large storage capacities, and/or private vehicles in homes aggregated by energy service companies in utility regions where a critical mass of vehicles is present.

Use of this storage capacity will likely align with peak demand periods when electricity prices are higher [38]—typically in the early evening. The coincidence of this period with the return home from work may inform the design of the service agreements as outlined previously. Although there is much variation within and across markets, one study that drew on peak power costs for California suggested that the business case for V2G is marginal [38].

While a generalized approach delivered under a service agreement aimed at peak power applications may enhance V2G viability, the outcome will by necessity underutilize the potential grid support of individual vehicles. Conversely, V2G within a smaller, closed system under the management of a single entity may provide greater benefits for a similar investment in the enabling technology.

Corporate fleet vehicles operating out of owner-occupied premises [39, 40] or private vehicles garaged at home may provide building energy support and improved vehicle asset utilization through user coordination across their electricity and transport needs [41]. This scenario becomes increasingly relevant when the objective is to enable larger amounts of renewable energy to be integrated—refer to Sect. 4.5 below.

Finally, a comparison with stationary storage solutions is important in any consideration of the path forwards for V2G. Purpose-built grid storage facilities will provide scale, reliable access, and utilize battery designs optimized for the task (for example, flow batteries are an emerging technology particularly well-suited to grid capacity storage) [42]. While the use of plug-in vehicles for storage improves utilization of an existing asset, preferential allocation of vehicles to transport tasks may undermine their competitive position for grid support [43].

The obvious solution may therefore be a complementary arrangement between stationary energy storage facilities and electric vehicles [44]. Stationary facilities may provide the main form of grid storage that is in relatively constant use, while vehicles may be called upon for only the highest value/least frequent storage applications—this is partly the thinking behind ancillary services applications explained in Sect. 4.6. This approach recognizes the contrasting characteristics of the storage technologies, the lower unit power supply costs of stationary storage, and the opportunity to realize additional value from investment in vehicles.

4.5 Grid Storage for Renewables Integration

Affordable energy storage is increasingly viewed as the key enabler for large-scale renewable energy [45]. The intermittent, variable supply characteristics of renewable energy resources create significant challenges for their integration into the grid and widespread adoption. These obstacles may be surmounted by storing energy during periods of oversupply for later re-use or to provide compensation for power quality impacts (which are dealt with as ancillary service applications in Sect. 4.6 below).

As has been described previously, the main limitation to this vision lay in the competing uses of the vehicle. Patterns of vehicle use are unrelated to those of renewable energy generation, suggesting that the requirements for renewable energy storage may be often incompatible with the availability of stationary vehicles. This obstacle may be partly countered through optimized charging strategies [20] that take advantage of EVs to store surplus renewable energy for use in transport, rather than supply back into the grid (refer also Sect. 4.2).

At the macro scale the continued uptake of both technologies will allow the variability of individual resources to be accommodated and taken advantage of across a regional network [46]. As more EVs are adopted and made available as a supplementary storage resource alongside stationary facilities, the growing and flexible storage capacity will facilitate larger amounts of renewables [45]. Coordination of the ever-expanding number of complex grid resources will be handled by the continued evolution of Smart Grid control systems.

At the other end of the scale, the benefits of allying the technologies within smaller-scale systems at the precinct, campus, building or home level can be captured by a single entity. This may encourage adoption due to direct and indirect financial, social and environmental benefits [47].

For example, the optimized charging strategy touched on above may allow system operators to maximize the benefits of their renewable energy production. This outcome will reflect a range of factors including purchased electricity costs, feed-in tariffs, and conventional transport fuel costs [25, 48].

Careful accounting and reconciliation of energy production and use may also inform greenhouse gas mitigation strategies, and by extension brand-building efforts in the eyes of key stakeholders. Technology providers have already begun to exploit this vision as a means of promoting their products [49].

As an extension of this, the allure of "going off the grid" should not be underestimated [50]. With solar generation making the transition to mainstream adoption in many markets, these homeowners are engaged in their energy use and receptive to the integration of an EV into their home [51]. Early-adopters will pay a premium in terms of cost and effort to attain a technology-rich, low-carbon lifestyle which is technically possible right now (particularly when coupled with a stationary storage solution). Less-costly, more user-friendly solutions will have increasing appeal in the face of rising network connection costs that are unfolding in many markets.

4.6 Grid Storage for Ancillary Services

Grid operators use ancillary services to maintain stability and reliability in the face of dips and surges in the balance of electricity supply and demand [23]. There are a range of ancillary service products that reflect the varying timescales over which this response is required, including frequency response, spinning reserve, regulation and load following/ramping. Due to the specific and critical needs addressed by ancillary services, their value tends to be significantly higher than equivalent amounts of energy supplied as capacity. Payments for ancillary services include payments for availability and for delivery [52].

Electric vehicles as grid storage resources are potentially well-suited to ancillary services that require fast response times and are of short duration [53]. These requirements are a good match for EV storage technology characteristics, and are less constrained by the limitations on EV storage volumes and depths of discharge.

Participation in ancillary services markets, if permitted under the market rules, would likely take place via an energy aggregator such that the capacity of multiple vehicles could be pooled and traded into the market. Opinion on the business case for EVs as a resource within ancillary services markets varies [38, 53].

5 The Path Forwards

Establishing a roadmap for EVs as grid support begins with an understanding of the electricity market. The rules and regulations applicable to grid and market operation are a threshold issue for the use of EVs as grid support. They set out what is permitted, very often determine the associated financial value, and strongly influence the distribution of costs and benefits across the value-chain.

As was outlined in Sect. 2, market design and operation varies from region to region. This has implications for technology and service providers, as the market opportunity must be significant enough to warrant their investment. The path forwards for EVs as grid support is likely therefore to reflect the arrangements and market-forces prevailing across the largest markets for the vehicles.

On this basis the near-term trends will be likely determined by the U.S. market, particularly California (the region with the highest plug-in sales globally, the operation of which takes place within the Californian electricity market) [54]. As the Japanese and EU markets receive largely the same plug-in vehicle makes and models as the U.S. (unlike China), issues and opportunities common to all three markets will also be a factor in the development plans of technology providers.

Efforts to address peak electricity demand are translating to consideration or application of Time-of-Use (TOU) tariffs in these markets. Cars are generally supplied with the capability to manage charging so that drivers can take advantage of these tariffs easily and conveniently. Although there are barriers to the adoption of TOU tariffs in markets where they have not already done so [55], the low costs and high value associated with the promotion of off-peak charging for EVs is likely to translate to alignment with EV uptake.

Some U.S. utility regions are also operating direct load control as an option for a number of electrical loads. Although no markets currently incentivize or operate this for EV charging management, in November 2013 the Californian electricity market regulator initiated rule-making to examine "the potential value to ratepayers and the electric grid of vehicle-grid integration, including the use of vehicle batteries for Demand Response or energy storage" [56]. This follows on from the

Californian mandate for 1.3 GW of batteries and grid storage by 2020 [57]. When combined with their market leadership in sales of plug-in vehicles, the effect of these initiatives will be to ensure that California is the lead market for EVs as grid support.

Should it transpire that there is a business case in favour of optimized charging and/or V2G interactions of whatever type, the likely pathway would be via third-party energy service providers experienced in the delivery of Demand Response programmes. These entities will aggregate vehicles [58] under service agreements that allow the service provider to participate in the wholesale electricity market. Ensuring a fair distribution of costs and benefits will require solutions to be found for the upfront costs of the technology enablers in each application. The findings of the Californian regulator from their investigation of vehicle-grid integration will inform any response which may address this issue and in doing so advance the market. Notably, optimized charging may provide a superior cost-benefit argument with less barriers to adoption than vehicles-as-storage applications [38, 53].

As behind-the-meter interactions are less influenced by the diverse, complex and slow-to-change electricity market environment, these applications may quickly surpass those revolving around market participation.

Vehicle-to-Home (V2H) technology progressed rapidly in Japan as an outcome from their response to the Sendai earthquake and associated events of 2011. In 2012 vehicle manufacturers released systems into the Japanese market that allow vehicles to be used for emergency back-up, such that the Nissan LEAF-to-Home system is now bought by around 7 % of Japanese LEAF drivers [59]. Although Japanese and North American electricity systems both utilize 110 V supply, vehicle manufacturers are yet to offer back-up systems for sale in the U.S. The availability of aftermarket EV back-up power devices in the U.S. suggests that this oversight may be temporary.

The emergence and continued evolution of the "Connected Home" combined with the growth in residential solar installations is likely to have consequences for V2H applications. Clean technology enthusiasts are likely to form the V2H early-adopter market who will pay a premium to align themselves with the powerful vision symbolized by these systems. To capture the early-mover advantages for their brands, V2H system providers may emerge fairly rapidly from the Japanese experiences with emergency back-up systems and/or strategic partnerships between Electric Vehicle and Connected Home technology vendors. These opportunities will combine with the convergence towards recognized technical standards (particularly for the Connected Home) to help accelerate development of V2H products.

At the nexus of regional-scale V2G applications and V2H sit Vehicle-to-Building (V2B) interactions within commercial/industrial applications. Although many of the V2H benefits above may be harnessed by corporates, the obstacles inherent to tenanted premises and the disconnect between the fleet/facilities managers and the actual drivers of the vehicles are a barrier to uptake. For corporates who operate plug-in fleet vehicles out of owner-occupied premises, near-term brand-building opportunities exist for V2B applications, even if these are likely to be scaled to a small number of vehicles. As the potential marketing benefits decrease with time, the business case for V2B against on-site stationary storage solutions must overcome the clear disadvantages relating to management of the combined functions of the vehicle as transport and storage.

6 Conclusion

Convergence of our electricity and transport systems has never seemed closer. The dream of many to become self-sufficient for their daily energy production and use seems more attainable than ever. These views are founded on the observation that technical solutions exist right now to allow for electric vehicles to be fully-grid-integrated.

The commercial reality is however more sobering. This paper has provided an explanation of the challenges to realization of the "off-grid" vision. Regulatory and financial obstacles must be addressed before electric vehicles will become a significant contributor to the Smart Grid.

Opportunities do however exist for a range of niche applications, such as back-up energy supply during times of emergency, along with optimized charging practices that will provide more widespread benefits. Through lessons learned along the way and the continued evolution of more affordable, effective technology, the ultimate goal for stationary vehicles to provide cost-effective grid storage may be realized.

References

1. Productivity Commission (2013) Electricity network regulatory frameworks—inquiry report vol 1. Chapter 9: demand, 9 Apr 2013. http://goo.gl/dYluy. Accessed 13 July 2013
2. Clearly Energy (2014) Residential demand response programmes. http://goo.gl/iVQVKp. Accessed 22 Jan 2014
3. Handberg K (2013) Victorian electric vehicle trial mid-term report, Victorian government, June 2013. http://goo.gl/QrJNE. Accessed 20 Jan 2014
4. Schey S (2013) The EV project: Q2 2013 report, ECOtality North America report for the U.S. Department of Energy, Aug 2013. http://goo.gl/NnYR6M. Accessed 20 Jan 2014
5. GM-Chevrolet (2014) Volt specifications. http://goo.gl/y2DU1A. Accessed 16 Jan 2014
6. Nissan USA (2014) LEAF specifications. http://goo.gl/6iNZQV. Accessed 16 Jan 2014
7. Tesla Motors (2014) Tesla model S battery. http://goo.gl/hraQ64. Accessed 15 Jan 2014
8. BYD (2014) Electric bus. http://goo.gl/hN8sgX. Accessed 16 Jan 2014
9. Energy Information Administration (2014) How much electricity does an American home use? http://goo.gl/Y81aTJ. Accessed 15 Jan 2014
10. Blanco S (2014) Chevy volt replacement battery cost varies wildly, up to $34,000, Autoblog Green, 10 Jan 2014. http://goo.gl/N83qdH. Accessed 16 Jan 2014
11. Handberg K (2012) Environmental impacts of electric vehicles in Victoria, Victorian Government, Nov 2012. http://goo.gl/7ECyVp. Accessed 20 Jan 2014
12. CHAdeMO, *CHAdeMO* chargers (2014). http://goo.gl/1oAuYz. Accessed 16 Jan 2014

13. Nissan Motor Company (2014) LEAF to home electricity supply system. http://goo.gl/J4i9VE. Accessed 24 Jan 2014
14. Dick R (2013) SPIDERS delivers first-of-a-kind bi-directional electric vehicle chargers at Fort Carlson, Colorado, Burns & McDonnell media release, Marketwired, 30 Aug 2013. http://goo.gl/kCJWbb. Accessed 24 Jan 2014
15. Briones A, Francfort J, Heitman P, Schey S, Schey S, Smart J (2012) Vehicle-to-grid (V2G) power flow regulations and building codes review by the AVTA, Idaho National Laboratory report for the U.S. Department of Energy, Sept 2012
16. Powerup Consortium (2011) About the powerup project, 9 Mar 2011. http://www.power-up.org/?p=1. Accessed 24 Jan 2014
17. De Hoog J, Handberg K, Jegatheesan R (2013) Demonstrating demand management: how intelligent EV charging can benefit everyone. EVS27 conference paper, Nov 2013. http://goo.gl/pUzQJN. Accessed 20 Jan 2014
18. Schey S (2013) The EV project: Q2 2013 report, ECOtality North America report for the U.S. Department of Energy, Aug 2013. http://goo.gl/NnYR6M. Accessed 20 Jan 2014
19. Massachusetts Institute of Technology (MIT) (2011) The future of the electric grid, MIT, 5 Dec 2011. http://goo.gl/6bOuBx. Accessed 23 Jan 2014
20. Weis A, Jaramillo P, Michalek J (2014) Estimating the potential of controlled plug-in electric vehicle charging to reduce operational and capacity expansion costs for electric power systems with high wind penetration. Appl Energy 115:190–204
21. Eurelectric, Deploying publicly accessible charging infrastructure for electric vehicles: how to organise the market? A Eurelectric concept paper, July 2013. http://goo.gl/YVML7P. Accessed 22 Jan 2014
22. Handberg K, Angelovski Z (2013) Demand management of electric vehicle charging using Victoria's smart grid, May 2013. http://goo.gl/CT7hTK. Accessed 21 Jan 2014
23. Perlstein B, Gilbert E, Stern F, Corfee K, Batternberg L, Maslowski R, Schare S, Firestone R (2012) Potential role of demand response resources in maintaining grid stability and integrating variable renewable energy under California's 33 %. Renewable portfolio standard, navigant research white paper prepared for California's demand response measurement and evaluation committee, 20 July 2012. http://goo.gl/FjNoGF. Accessed 24 Jan 2014
24. Finn P, Fitzpatrick C, Connolly D (2012) Demand side management of electric car charging: benefits for consumer and grid. Energy 42(1):358–363
25. Tofel K (2013) Adding an electric car cut the payback point of our solar panel investment in half, Gigaom, 12 May 2013. http://goo.gl/RDzOl0. Accessed 21 Jan 2014
26. Mitsubishi Motors (2012) Mitsubishi motors to launch MiEV power BOX 1,500 W power feeder for its electric vehicles, 9 Mar 2012. http://goo.gl/hGHscG. Accessed 24 Jan 2014
27. Lavrinc D (2012) EV hack keeps homes humming after Hurricane Sandy, 8 Nov, Wired magazine. http://goo.gl/HWsfcex. Accessed 24 Jan 2014
28. National Fire Protection Authority (2013) NFPA 111: standard on stored electrical energy emergency and standby power systems, 2013. http://goo.gl/qQXZuf. Accessed 24 Jan 2014
29. Tri-County Electric (2013) Generator safety, 2013. http://goo.gl/y31lOA. Accessed 24 Jan 2014
30. Lilienthal P (2012) HOMER Energy helps the City of San Diego design microgrids for emergency services, 8 Nov 2010, HOMER energy media release. http://goo.gl/V04JJy. Accessed 24 Jan 2014
31. Marnay C (2013) Challenges of supply—an evolving paradigm, IEEE power & energy magazine, Sept/Oct 2013. http://goo.gl/Ie8YEb. Accessed 24 Jan 2014
32. Powerup consortium (2011) About the powerup project, 9 Mar 2011. http://www.power-up.org/?p=1. Accessed 24 Jan 2014
33. Ustun TE, Ozansoy CR, Zayegh A (2013) Implementing vehicle-to-grid (V2G) technology with IEC 61850-7-420. IEEE Trans Smart Grid 4(2):1180–1187
34. Navigant Research (2013) Energy storage in commercial buildings, Feb 2013
35. Bojanczyk K (2013) Home energy management systems: vendors, technologies and opportunities, 2013–2017, GreenTech Media Research, 29 Aug 2013

36. Hidrue MK, Parsons GR, Kempton W, Gardner MP (2011) Can vehicle-to-grid (V2G) revenues improve market for electric vehicles? University of Delaware presentation to the international energy workshop, 7 July 2011. http://goo.gl/ZnAspQ. Accessed 24 Jan 2014
37. Marnay C, Chan T, DeForest N, Lai J, MacDonald J, Stadler M, Erdmann T, Hoheisel A, Muller M, Sabre S, Koch E, Lipkin P, Anderson RW, Gerber S, Reid E (2013) Los Angeles air force base vehicle to grid pilot project, paper to be presented to ECEEE 2013. Summer study on energy efficiency, 3–8 June 2013, Lawrence Berkeley National Laboratory. http://goo.gl/Drkh9c. Accessed 4 Feb 2014
38. Kempton W, Tomic J (2005) Vehicle-to-grid power fundamentals: calculating capacity and net revenue. J Power Sour 144:268–279
39. Stadler M, Mendes G, Marnay C, Megel O, Lai J (2011) Analysis of electric vehicle interconnection with commercial building microgrids. Paper presented to the UCLA smart grid thought leadership forum, Lawrence Berkeley National Laboratory, 6 Apr 2011. http://goo.gl/ryxE1L. Accessed 4 Feb 2014
40. Tomic J, Kempton W (2007) Using fleets of electric-drive vehicles for grid support. J Power Sour 168:459–468
41. Mendes G, Stadler M, Marnay C, Ferrao P, Ioakimidis C (2012) Modeling of plug-in electric vehicles' interactions with a sustainable community grid in the Azores. Paper presented to 2012 ACEEE. Summer study on energy efficiency in buildings, 12–17 Aug 2012, Lawrence Berkeley National Laboratory. http://goo.gl/Hz8TAq. Accessed 4 Feb 2014
42. Office of Electricity Delivery & Energy Reliability (2013) Grid energy storage, United States Department of Energy, Dec 2013. http://goo.gl/QuZEkP. Accessed 5 Feb 2014
43. Stadler M, Marnay C, Kloess M, Cardoso G, Mendes G, Siddiqui A, Sharma R, Megel O, Lai J (2012) Optimal planning and operation of smart grids with electric vehicle Interconnection. Paper to be published in the journal of energy engineering, 2 Jan 2012, Lawrence Berkeley National Laboratory. http://goo.gl/uZcyI0. Accessed 4 Feb 2014
44. Khambadkone AM (2013) Energy storage in microgrids: comparing different energy storage technologies for microgrids in different geographical scenarios. Presentation to microgrid forum Asia, 11–13 Nov 2013, Singapore
45. Inage S (2009) Prospects for large-scale energy storage in decarbonised power grids. International Energy Agency Working Paper. http://goo.gl/NKdJhP. Accessed 5 Feb 2014
46. Proteus A (2011) Electric vehicles, smart grid, residential solar power: a virtuous circle emerges, CleanTechnica, 12 Sept 2011. http://goo.gl/0bs2TT. Accessed 5 Feb 2014
47. Anscombe N (2013) Green electricity drives demand for plug-in vehicles, Environmental Research Web, 11 June 2013. http://goo.gl/mdJqQL. Accessed 5 Feb 2014
48. Shahan Z (2011) Combining solar power and your electric vehicle—too practical to pass up (reader comment), 17 Sept 2011. http://goo.gl/4iAHlq. Accessed 5 Feb 2014
49. MyEnergi Lifestyle (2014). http://myenergilifestyle.com. Accessed 5 Feb 2014
50. Rosen N (2010) Off the grid: inside the movement for more space, less government, and true independence in modern America. Penguin Books, New York, July 2010
51. Voelcker J (2013) Electric cars vs solar panels: which one's the gateway drug? Green car reports, 4 June 2013. http://goo.gl/7pHsS9. Accessed 5 Feb 2014
52. Australian Energy Market Operator (2014) Ancillary services. http://goo.gl/K6Sykw. Accessed 5 Feb 2014
53. Leo M, Kavi K, Anders H, Moss B (2014) Ancillary service revenue opportunities for electric vehicles via demand response, project report for master of science from the school of natural resources and environment, University of Michigan, Apr 2011. http://goo.gl/jdCUfz. Accessed 5 Feb 2014
54. Implementing agreement for cooperation on hybrid and electric vehicle technologies and programmemes, hybrid and electric vehicles: the electric drive gains traction, International Energy Agency, May 2013. http://goo.gl/jzYdD8. Accessed 6 Feb 2014
55. Cooke D (2011) Empowering customer choice in electricity markets, information paper, International Energy Agency, Oct 2011. http://goo.gl/GgAAtk. Accessed 6 Feb 2014

56. Californian Public Utilities Commission (2013) CPUC continues work toward expanding the use of alternative-fuel vehicles in California, Press Release, 14 Nov 2013. http://goo.gl/q4ZdLD. Accessed 6 Feb 2014
57. State of California (2014) Bill No. AB 2514 energy storage systems. http://goo.gl/UqJgcu. Accessed 6 Feb 2014
58. Schmid D (2014) SwRI develops first ERCOT-qualified vehicle-to-grid aggregation system, media release by the Southwest Research Institute, 14 Jan 2014. http://goo.gl/r751F7. Accessed 10 Feb 2014
59. Shirakawa T (2013) Keynote presentation to EVS27. Nissan Technical Center Europe, Barcelona, 18 Nov 2013

Energy Efficiency in Electric and Plug-in Hybrid Electric Vehicles and Its Impact on Total Cost of Ownership

Matteo Conti, Richard Kotter and Ghanim Putrus

Abstract There is an increasing awareness, policies and an incentive landscape, which are encouraging and starting to shape future transport as part of a wider ecosystem of infrastructure, use, behaviour and sustainability. However, one of the main barriers for the wider uptake by both fleet and individual users of electric vehicles is the concern of the uncertainties of Total Cost of Ownership (TCO). This contribution is based on a mix of original modelling, simulation and laboratory experimentation studies as well as a review of the academic and policy literature. It focuses on vehicle design and the battery and energy management in electric and plug-in hybrid electric vehicles (EV/PHEV). EV users express concerns about the longevity of the electric battery and hence the life cycle (especially with frequent fast charging), which amounts to a major part of the costs and value of the vehicle. Using the battery to provide ancillary services will add more value to the EV and reduce the effective TCO.

Keywords Electric vehicles · Energy efficiency · Total cost of ownership · Vehicle design · Battery · Smart charging

1 Introduction

The growing requirement to produce more energy efficient ultra-low carbon vehicles (ULCVs) represents a major technical and financial challenge for major vehicle manufacturers and component suppliers. Some of the current limitations of battery

M. Conti (✉)
Royal College of Art, Vehicle Design, Kensington Gore, London SW7 2EU, UK
e-mail: matteo.conti@rca.ac.uk

R. Kotter · G. Putrus
Northumbria University, School of Engineering and Environment, Ellison Place, NE1 8ST, Newcastle upon Tyne, UK
e-mail: richard.kotter@northumbria.ac.uk

G. Putrus
e-mail: ghanim.putrus@northumbria.ac.uk

D. Beeton and G. Meyer (eds.), *Electric Vehicle Business Models*,
Lecture Notes in Mobility, DOI 10.1007/978-3-319-12244-1_9

technology in terms of energy density, power density, weight and cost [1] have led automakers to focus not only on the development of electric and hybrid powertrains but also on the optimization of other critical vehicle technical areas aimed at enhancing the vehicle overall energy efficiency, such as reducing the weight and drag [2]. The use of on-board ancillary systems (heating, lights, etc.)—in addition to driving style, vehicle speed and the type of journey—are also key factors to improve the vehicle energy efficiency, range and performance [3].

This paper provides an overview of the main vehicle design advances to increase Electric Vehicles (EVs) and Plug in Hybrid EVs (PHEVs) market penetration by offering vehicles with an increased range and consequently becoming a more attractive business proposition to prospective buyers or users. Some key technical constraints are also discussed as EV and PHEV technology is not yet at a mature stage [4]. In addition, an account of the design and operation of the battery, smart charging, eco-driving, vehicle energy consumption and management are provided. Finally, this paper presents an evaluation, based on modelling, simulation and laboratory experimentation studies, on how these may be designed to lower the total cost of ownership (TCO), since this is critical for take-up.

2 EV/PHEV Design to Reduce Energy Demand During Driving Conditions

2.1 EV/PHEV Body Aerodynamics

Aerodynamics play a prominent role in the overall efficiency of a vehicle. Most of the EVs and PHEVs available on the market feature an optimised aero package specifically designed to lower their *drag coefficient* (Cd) caused by airflow turbulence. It is common for vehicles such as the Toyota iQ EV to feature aerodynamic enhancements applied to the front grille opening and body panels to separate and improve air flow, as well as underfloor covers to minimise turbulence [5].

Similarly, the 2013 Fiat 500e low-volume EV is characterised by a reshaped front and rear end, revised wing mirror covers, small spats on the wheel wells, and under-trays to even out airflow [6, 7]. Aero improvements are also applied to some Internal Combustion Engine (ICE) vehicles in the shape of the special low carbon edition models, badged for instance *Blue efficiency* [8] by Mercedes-Benz, or *DRIVe* [9] by Volvo. From a comparative analysis, shown in Table 1, it follows that EVs and PHEVs are marginally more aerodynamically efficient than their ICE based counterparts.

This trend generally applies to the latest breed of EVs and PHEVs which have been designed as electrified ultra low carbon vehicles (ULCVs) right from the outset rather than being derivative versions of their existing ICE models. However, the latest BMW i3 and Nissan LEAF feature low drag bodies which are not significantly better than ICE vehicles of their respective size. The compared Cd figures

Table 1 Drag coefficient (Cd) comparison between internal combustion engine models and derived low carbon vehicle models [7–13]

Make	Model	Drive	Year	Cd	ΔCd	ΔCd%
Fiat	500	ICE	2007–present	0.36		
	500e	EV	2013–present	0.31	−0.05	−13
Honda	Mk7 Civic Sedan	ICE	2001–2005	0.30		
	Mk7 Civic Sedan	Hybrid	2003–2005	0.28	−0.02	−6
Scion (Toyota)	iQ	ICE	2008–present	0.31		
	iQ	EV	2012–present	0.31	0	0

Table 2 Low carbon vehicles drag coefficient (Cd) comparison [10, 14–18]

Make	Model	Drive	Year	Cd	ΔCd	ΔCd%
Honda	Mk1 Insight	Hybrid	2000–2006	0.25		
	Mk2 Insight	Hybrid	2010–present	0.28	+0.03	+12 %
Toyota	Mk2 Prius	Hybrid	2004–2009	0.26		
	Mk3 Prius	Hybrid/ PHEV	2010/2012–present	0.25	−0.01	−4 %
Nissan	Mk2 Note	ICE	2013–present	0.30		
BMW	i3	EV	2012–present	0.29	−0.01	−4 %
Mercedes Benz	Mk2 B Class	ICE	2012–present	0.26		
Nissan	LEAF	EV	2011–present	0.29	+0.02	+8 %

(as shown in Table 2) suggest that some of the latest ICE models perform surprisingly well against the most evolved EVs available on the current market.

Although there are some after-market PHEV aero body kits such as the Aero Prius YuraStyle [19, 20], for the Toyota Prius, those designs are too extreme in terms of styling to be accepted by the vast majority of Prius owners [18]. Allegedly their improved Cd figures, obtained by further covering of the car rear end and wheels [21], may provide some tangible gain in terms of the vehicle extended range [20, 22] compared with the standard PHEV model, driven in everyday variable driving conditions.

2.2 EV/PHEV Kerb Weight

Air friction is not the only key parameter responsible for reducing vehicle range. The overall car kerb weight is another major technical challenge for auto makers as they aim to lower the vehicle mass as much as possible.

The lithium-ion traction battery in the Tesla Roadster weighs about 453 kg [23]. However, designers managed to offset the battery weight with the adoption of a full aluminium frame and plastic body panels to keep the EV car weight down to

1,220 kg [24]. This valid technical approach normally proves to be more costly when it is applied to conventional ULCVs designed for daily use.

However, there are different approaches to increasing the vehicle range and efficiency of ULCVs. Chevrolet managed to increase the range of its Volt PHEV by reducing its aerodynamic drag which can be advantageous when regaining energy through regenerative braking [25]. BMW adopted a radical design of the i3, which is the first mass-produced vehicle in the world to feature a carbon-fibre reinforced plastic (CFRP) body structure [26, 27]. It is likely that future LCVs will also make use of this new chassis construction method as a common platform onto which an aluminium frame is mounted to house the battery and powertrain.

In the typical daily use of an EV/PHEV, predominantly in an urban environment, where low speed limits are enforced, the vehicle gross weight counts more than its drag. In simple terms energy calculations can be easily deducted from the following Newton's inertial force equation [28]:

$$F = ma \tag{1}$$

where F is the Force needed to move the ULCV; m is the mass of the ULCV; and a the vehicle acceleration. If the mass value increases more force will be then required to obtain a given acceleration. This is why urban driving is less energy efficient than driving on motor ways, where acceleration is reasonably constant. In a different driving situation, e.g. an ULCV driving up a gradient is heavily affected by its weight, therefore reducing its driving range. For the above-mentioned conditions a lightweight ULCV is preferable. On the contrary, the increase in weight has its benefits after the ULCV acquires a certain speed, which is being kept reasonably constant (e.g. driving on a motorway) as the vehicle will carry more momentum (kinetic energy) to move along.

In the case of an ULCV used primarily for journeys beyond the city boundaries, at higher speeds, typically above 50 mph, the vehicle air drag begins to affect the amount of power required to propel the vehicle forward and maintain or increase its speed. This is due to the fact that [29]:

$$F = \frac{1}{2}\, \rho\, v^2\, Cd\, A \tag{2}$$

where F is the drag force needed to move the ULCV; ρ is the air density, v the velocity, Cd is the coefficient of drag, and A is the vehicle cross sectional area.

Assuming that ρ, Cd and A are constant (a specific vehicle), the change in the drag force (F) that results from a small change in vehicle velocity can be derived as:

$$\frac{\Delta F}{\Delta v} \approx \frac{\partial F}{\partial v} = 2\,(\frac{1}{2}\, \rho\, v\, Cd\, A) = 2\left(\frac{F}{v}\right) \tag{3}$$

which gives:

$$\frac{\Delta F}{F} \approx 2\frac{\Delta v}{v} \tag{4}$$

That is, a relative change in vehicle speed results in twice the relative change in the drag force needed to move the vehicle. This means that every time the vehicle speed doubles, the drag force value quadruples [29] as the amount of energy (Wh/mile) absorbed in aerodynamic losses. The other variable that greatly affects the ULCV range is the Cd, which depends on A. Thus, a low drag ULCV is the better option. With existing EV/PHEV battery technology, the increase in battery capacity, to extend the EV range, results in a larger battery mass and heavier vehicle [30].

It is foreseeable that in the future EVs/PHEVs will feature more sophisticated on-board control which will be able to optimize the ULCV efficiency and range using satellite navigation and maps to predict the vehicle journey continuously taking into account all the factors mentioned above. The ability to inform the driver of the vehicle's real time consumption also taking into account the wind speed and direction through constant updates from the Met Office, for instance, are part of the overall energy optimization which will be likely to be introduced in future ULCVs. Even apparently negligible power gains still have a significant summative effect on the overall performance of the vehicle.

2.3 EV/PHEV On-Board or Off-Board Charging

Today's battery technology constraints have led automakers to produce the majority of EVs with a limited range similar to the pioneering EVs produced in 1910 [31]. This, in addition to cost and charging which still presents a number of obstacles [32], are the main factors limiting the proliferation of EVs and PHEVs.

On-board charging enables EV/PHEV users to plug their cars in any domestic socket [33] or a power supply in various locations that are not necessarily in the proximity of a public charging point. However, Level 1 AC charging, running on 110 V in the US, means that a PHEV/EV could take from 6+ (for a PHEV) to 24+ h (for an EV) to charge [34]. The answer would be to upgrade to a Level 2 AC charging station, employing 220 V supply, installed at home or nearby, to allow a PHEV to be charged from 2+ to 8+ h for an EV [34]. This facility is also appropriate for charging vehicles at work, around shopping centres and supermarkets.

The very latest edition of the Nissan LEAF makes on-board charging more compatible with today's frantic lifestyle as its battery can be refilled within 4 h through the adoption of a dedicated 6.6 kW charger option powered by 32 A circuit [35]. In order to further reduce 'filling up' times, Level 2 DC charging has been introduced which uses an off-board charger. The charger is rated between 20 and 80 kW, which gives 3–5 miles' range for every minute of charge [34]. Fast chargers are expensive to install and therefore are not used for domestic charging; rather, they are popular for public and commercial charging stations.

Another viable solution to facilitate the use of EVs is represented by the concept of swapping the battery at a charging station, though the recent financial collapse of BetterPlace appears to be related to too low take-up, and too high costs [36].

3 Demand from Auxiliary (Non Power Train) Loads/Functions

Ancillary systems constitute an additional power load on the running of an EV and PHEV [37] which can significantly affect the range of the vehicle in various operating conditions. Typical functions such as climate control [heat ventilation, air conditioning (AC)], lights, info entertainment (radio, CD player, centre console display, satellite navigation, etc.) can reduce the PHEV/EV travel range. The battery management system (*BMS*) and other primary auxiliary circuits (instruments binnacle electronics, central locking, electric windows, immobilizer, etc.) used to govern vehicular operations, add an imperceptible energy demand. This auxiliary power demand (*PAux. Load*) may be represented as:

$$PAux.\,Load = P\,AC + P\,Lights + P\,Info\,Entert. + P\,BMS + P\,Aux.\,Circuits \quad (5)$$

The first three variables are particularly relevant to an EV as its driving range depends only on the main battery pack whereas a PHEV rely on an ICE unit to extend its range. An electric air conditioner with a peak load of 3 kW can reduce the EV range by 16–38 % depending on the driving cycle [38]. Its power absorption may vary between 0.2 and 2.2 kW. To compensate for such a power drain the latest 2013 LEAF adopts a new heat pump-based heating system, which is designed to use considerably less battery power [39].

The combined effect of internal and external lights is about 80 W [37] as efficient LED lights, used for turning signals, daytime running lights or brake lamps [40] are increasingly replacing traditional automotive bulbs and are commonly coupled with halogen and xenon headlights. In terms of audio equipment, manufacturers try to considerably reduce the auxiliary power consumption [41]; for example, the 2013 Nissan LEAF features a new premium Bose Energy Efficient audio [42] which uses about half of the electrical current of standard systems [43] whereas the Toyota Prius is fitted with a 120 W audio system that is comparable to 600 W systems [42]. The power consumption of BMS units, which consume on average between 15 and 40 mA (3–8 W) [44, 45] and all other auxiliary electrical systems is negligible.

The overall power consumption of EV ancillary systems is 10–33 % of the traction battery power, depending on the driver's choice and use of heat or air conditioning [46]. To compensate for such a load, an EV may be equipped with photovoltaic panels, where a 200 W system returns about 1 km of electric range for each hour of full direct sun exposition [47].

Future LCVs will be adopting a 42 V electrical system standard [48] in order to save weight and cost of the electrical components and improve energy efficiency. Manufacturers like Audi are planning to implement high voltage technology on 48 V mild-hybrid platform [49] in the short term.

4 Battery Cycle Life and State of Health

The battery is the most expensive part of an EV and accounts for over 50 % of the total production costs of the vehicle [50]. Current EVs/PHEVs are usually powered by high capacity Lithium-ion batteries, which vary in size from a few kWh to few tens of kWh capacity. The battery capacity determines the vehicle electric range and, with current battery technologies, this is limited to around 200 km or less for commercial EVs/PHEVs. These two factors (cost and range) emphasize the importance of maintaining the battery in a healthy state for as long as practically possible in order to reduce the TCO.

Another area that will help reduce the TCO of EVs/PHEVs (with market growth and high deployment) is their use to support the grid. This can be in their use as a controllable load during charging or as storage in what is called Vehicle to Grid (V2G), where the EV could be used to store surplus output from renewable generation and release this back to the grid during peak demand [51, 52]. Such application requires appropriate control and communication with the grid, vehicle user and battery management system [51] and may be implemented as part of the 'smart grid' concept [53]. However, providing this service by the EV/BHEV has negative impact on the battery cycle life and consequently on the vehicle TCO.

Battery State of Health (SOH) is defined as the difference between the usable capacity and the end of life capacity and is usually expressed as a percentage of the rated capacity [54]. EV manufacturers define the end of life capacity of the battery as the state when the battery capacity reaches 80 % of the rated (fresh) capacity [54]. Therefore, it is important to understand the factors that affect battery degradation and provide the means to optimize battery SOH, not only reduce the TCO but also conserve vehicle range. Battery performance depends not only on the battery chemistry but also on external factors, such as surrounding temperature and the way the battery is being used. Capacity loss in Lithium-ion batteries may be attributed to two reasons: "calendar life" loss and "cycle life" loss.

The calendar life is the continuous slow degradation of the battery due to the passage of time, whether the battery is being used or not. It is largely affected by the storage temperature and the charge state. Extreme ambient temperatures and high average State of Charge (SOC) result in fast degradation. This type of degradation can be attributed to permanent chemical change and thus follows Arrhenius law [55]:

$$\frac{dC}{dT} = Ae^{kT} \tag{6}$$

where T is the absolute temperature, C is the battery capacity. A and k are numerical constants that depend on battery chemistry and are usually determined by experimental tests. The cycle life depends on the chemistry of the battery as well as the way the battery is being used during charging and discharging. It is affected and hence determined by four main factors which are interlinked. These are: the charging/discharging current rate, battery temperature, SOC and depth of discharge (DOD). The significance of each parameter and its impact on the cycle life is usually estimated through experimental cycling tests, with varying accuracy. Usually, degradation caused by cycling is much faster than that caused by calendar loss (storage), but obviously this depends on the EV (battery) usage.

Test results show that battery cycle life drops with increased charging/discharging current rates, e.g. if the current rate goes up from 0.74 to 1 C, the battery cycle life drops from around 1,000 to around 200 [56]. Test results demonstrate that Lithium ion batteries perform best and have a longer cycle life at room temperature around 20–25 °C [57]. They also perform better at low (less than 50 %) average SOC. For example, the battery cycle lifetime when cycled at 15 % SOC is over 3 times the cycle lifetime with a SOC of 95 % [58]. Test results also show that at fixed temperature and current rate, the battery cycle life decreases with increasing DOD. For example, if battery capacity is fully used (100 % DOD), the battery capacity drops to 80 % of its initial value after 1,600 cycles. When the DOD is 30 % or less, the battery cycle life increases significantly to 10,000 times [59].

Since different parameters that affect battery degradation (both calendar and cycle lives) are interlinked, it is difficult to exactly quantify the individual impact of these parameters. However, test data available provide valuable insight into the impacts and therefore help in the design of battery management systems and charge/discharge controllers that will optimize battery SOH.

As mentioned earlier in this section, battery degradation also depends on battery technology (chemistry) and this is continually improving and, together with appropriate control, can help in prolonging battery life and reducing the TCO of the EV/PHEV.

5 Smart Battery Charging

Smart charging is a crucial element in the realization of a safe, adaptable and sustainable power network which is able to cope with an increasing numbers of EVs and PHEVs creating an additional energy load on the grid. An adequate control and management of charging is necessary to avoid poor power quality and possible electricity supply failures which can occur with high penetration of EVs [60]. Research has shown that network voltage levels may deviate from the statutory limits even with small penetration levels, say 10 %, in weak parts of the distribution networks [61]. In addition, charging at peak demand on the grid would result in higher CO_2 emissions and electricity rates [60].

The adoption of smart charge controllers that initiate and stop charging in response to the conditions of the power grid [62] can minimise the impact of charging on the grid and minimise electricity expenses for EV or PHEV users. Smart charging may be designed to optimize EV and PHEV battery State of Charge (SOC) and calculate in real time appropriate charging patterns based on battery state of health and local power distribution voltage [62], taking into account the user's request. This complex operation can only be performed by a smart controller which also stabilises the grid by monitoring the incoming AC voltage, and frequency. In order to meet EVs and PHEVs deployment targets governments need to develop a concurrent network of rapid charging facilities and ensure that the energy network providers involved are fit for purpose when it comes to coordinating their services. By 2020, countries that are members of the Clean Energy Ministerial (CEM) Electric Vehicle Initiative (EVI) have set cumulative targets to install about 2.4 million slow chargers and 6,000 fast chargers [62]. This infrastructure expansion through the mass adoption of smart controllers provides an opportunity to incentivise customers with different tariffs throughout the day and influence their charging behaviour whilst maintaining the grid power management dynamic and balanced.

Energy generation in the future will rely more on renewables, where customers will decide to invest in PV panels, wind turbines or other low carbon generation technologies to produce part or all of their domestic electricity needs. This scenario will become reality, when the next generation of EVs, fitted with more powerful batteries, will be introduced to coexist and exploit the use of small-scale electricity generation from low carbon technologies. These vehicles will be capable of storing power to assist the grid balance and stability which becomes a problem with high penetration of intermittent renewable power generation [63]. At the heart of this system is the smart charging controller, which will provide an active and reliable control to support the network operation (offset voltage sag and swell) and meet the EV user requirements [64]. These charging controllers will enable EVs' TCO to be further reduced whilst ensuring a satisfactory EV battery State of Health (SOH) and its durability.

6 Total Cost of Ownership (TCO) of EV/PHEV

Considering the increasing CO_2 reduction legislation currently implemented in Europe and North America [62, 65], there is an increasing demand for ultra low carbon 'green' cars [66, 67], which include EVs and PHEVs. However, the market penetration of EVs is still well below the forecast figures as the global EV stock represents only 0.02 % of all passenger cars [62] and their TCO remains high when compared with ICE based vehicles.

The financial drawback for EVs is constituted by the cost of the vehicle or its finance monthly payment, in addition to the battery lease monthly payment. This compound financial effect applies to all EVs available on the market today and it relates to the capacity of their traction battery.

In the following analysis, EVs are compared by considering their actual manufacturers' retail prices (ownership of vehicle and battery) whilst disregarding other relevant parameters such as the government subsidy or tax credit, vehicle standard equipment, technical refinement and brand name. The ratio of the vehicle cost to its maximum driving range is used to provide an indicative sense of the customer's EV value for money. According to Albert Lam, from Detroit Electric EV, batteries are responsible for about 54 % of the production costs of an EV, 26 % of the costs represent the drive system and the remaining 20 % is the car body manufacturing [50].

The lowest priced Nissan LEAF model, called Visia and featuring a 24 kWh battery, is offered at £26k in the UK, whereas the Tesla EV is sold at premium prices based on its large battery size. Considering the EV cost per mile it appears that there is a contained difference between the two vehicles examined (as shown in Table 3), although their respective TCO is greatly affected by the initial retail price.

The same principle applies to PHEVs which are currently on the market.

With the popular Toyota Prius very keenly priced, the Chevrolet Volt's TCO remains less favorable although it offers an extensive EV autonomy. The Volvo V-60 price bracket puts this executive PHEV and first diesel hybrid car in the world [73] in a different category altogether. The same can be said about its steep TCO. It can also be deduced that the Chevrolet Volt using a 16.5 kWh battery pack is about $5,000 more expensive than the Nissan LEAF, mostly due to its hybrid powertrain. In the case of the Toyota Prius costing below $30K it may seem that the sub $30K price bracket is necessary for those extended-range vehicles to go mainstream [74]. Table 4 offers a comparison of these three models.

The current TCO trend has been challenged by Chevrolet as its latest offering for the new 2014 Spark EV 1LT on a low-mileage lease is significant since it offers the most affordable EV on the market for $199 per month for 36 months with an initial

Table 3 EVs retail cost/range ratio comparison [68, 69]

Make	Model	Retail price	Market	Range (mile)	Cost/range ratio (per mile)
Nissan	LEAF Visia 24 kWh	£25,990	UK	124	£209.6
Tesla	Model S 60 kWh	£54,900	UK	240	£228.7

Table 4 PHEVs retail cost/range ratio comparison [70–72]

Make	Model	Retail price	Market	(EV range) Total range (mile)	Cost/total range ratio (per mile)
Chevrolet	Volt 16.5 kWh	£35,255	UK	(50) 300	£117.5
Toyota	Prius T3 4.4 kWh	£21,064	UK	(15) 540	£39
Volvo	V60 11.2 kWh	£48,670	UK	(31) 745	£65.3

deposit of \$999. This offers includes the full US federal tax credit which ranges from \$0 to \$7,500 [74]. In relation to the customer's TCO of an EV, it clearly appears that those vehicles with a limited range will be more appealing to the general public as their asking price will drop below \$20K [74].

In order to further reduce TCO, a viable alternative is to charge EV users per usage, based on the common and well-established mobile phones business model. Considering the EV battery second life, a 5-year buy back guarantee with residual value would reduce the initial battery cost.

A study conducted by Berger [75] forecasts that EVs' TCO may be competitive against ICEVs from the year 2015, based on a 3 year car lease with an annual mileage of 12k miles. The business model of leasing EV batteries is currently regarded as a means of reducing EVs' TCO for consumers as it offers an extended battery guarantee and reduced monthly costs.

7 EV/PHEV CO_2 Impact and Production Costs

When it comes to the production of vehicles, OEMs are committed to reducing CO_2 emissions in innovative ways. Responsible and more sustainable ULCV manufacturing is applied in the Chevrolet Volt assembly plant as it employs 516 kW PV panels [76] to reduce its dependence on the power grid. These facilities reuse, recycle or convert to energy all waste created in their daily operations, which conserves resources. In addition, the Volt ICE, transmission and battery facilities are landfill-free.

The BMW's €400 million i factory drastically reinvents and simplifies car mass production and use of resources. The i3 city car features only 100–120 separate parts in its body structure, compared with about 400 parts in a typical steel body [77]. To paint its plastic body panels, BMW introduced a new type of paint shop, which produces no wastewater and has a fifth of the normal cost for a paint finishing facility for steel-bodied cars. As a result this factory uses about 70 % less water and half the electricity [77] if compared with a common car plant.

The Mitsubishi Outlander PHEV uses green plastics, applied to high heat resistance areas, which are derived from the oils extracted from waste cashew nuts. This allows a reduction of CO_2 emission by up to 12 % compared to common petrol-based plastics [78]. The Outlander employs a high-capacity 12 kWh lithium-ion drive battery which enables an EV drive mode cruising range of over 55 km and an overall range in the region of 880 km [78]. Its CO_2 emissions figure is as low as 44 g/km when the battery is fully charged, but it reaches up to 135 g/Km when the battery is depleted [79]. These CO_2 emissions confirm the fact that in most long journeys PHEVs are not necessarily cleaner than modern ICEVs [79].

A study by the Union of Concerned Scientists in the US comparing the global warming emissions from EVs with those from gasoline-powered vehicles and found that: nationwide, EVs charged from the electricity grid produce lower global warming emissions than the average compact gasoline-powered vehicle (with a fuel economy of 27 miles per gallon)—even when the electricity is produced primarily from coal in regions with the "dirtiest" electricity grids; in regions with the "cleanest" electricity grids, EVs produce lower global warming emissions than even the most fuel-efficient hybrids; and EVs charged entirely from renewable sources like wind and solar power produce virtually no global warming emissions [80].

Other authors comparing conventional and alternative vehicle option from an environmental and economic perspective argue that if electricity is generated from renewable sources, the electric car offers advantages over the hybrid vehicle; but if the electricity is generated from fossil fuels, the electric car remains competitive only if the electricity is generated onboard. Yet if the electricity is generated with an efficiency of 50–60 % by a gas turbine engine connected to a high capacity battery and electric motor, the electric car is superior in many respects [81]. To charge EVs, studies indicate that the amount of generated CO_2 emissions per kilometre is between 52 and 70 g CO_2/km [23].

Plug-in hybrid electric vehicles (PHEVs) consume both gasoline and grid electricity. The corresponding temporal energy consumption and emission trends are valuable to investigate in order to fully understand the environmental benefits. The 24-h energy consumption and emission profile depends on different vehicle designs, driving, and charging scenarios. For example, a Californian scenario study assesses the potential energy impact of PHEVs by considering various charging scenarios defined by different charging power levels, locations, and charging time, with vehicle parameters based on realistic assumptions consistent with projected vehicle deployments. Results show that the reduction in petroleum consumption is significant compared to standard gasoline vehicles and the ability to operate on electricity alone is crucial to cold start emission reduction. The benefit of higher power charging on petroleum consumption is, however, small. Delayed and average charging are better than immediate charging for home, and non-home charging increases peak grid loads [81]. With rising fuel costs, EVs and PHEVs can be expected to deliver a certain level of financial benefit to consumers if seen on a longer time horizon of use, and depending on tax incentives and other public policy measures.

A Californian scenario study of charging demand shifts on an hourly basis for four diverse scenarios based on different electric circuit characteristics shows that circuit upgrades bring faster charging times, and reduce charging time differences between PHEV20 and PHEV60, with home charging replacing 40–50 % of distances currently travelled using ICEs with electric power for PHEV20 and 70–80 % for PHEV60. If charging facilities are available in public parking facilities, which will lead to more daytime charging, PHEV20 can convert 60–70 % of mileage from fuel to electricity, and 80–90 % for PHEV60. Emission reductions will be higher than those percentages since PHEVs will cover a greater fraction when measured by the number of trips, which emphasizes the equivalent number of ICE starts. The study concludes that it is not certain that diverting charging demands to off-peak periods will maximize energy efficiency, since daytime charging will allow more trips by electricity, but will result correspondingly in higher peaks for high-demand-periods. There are limitations to the assessments provided by this study—and many others—as it does not fully account for environmental impacts from PHEV penetration. Specifically, increased emissions and other types of energy usage regarding extra grid electricity demand are not assessed [82].

8 EV/PHEV New Business Models and TCO Reduction Contributions Across Industries and Regulatory Context

Although the cost of a traction battery, at $1,000 per kWh in 2008, has rapidly fallen to $485 per kWh in 2012 [64], it may take 3–4 years for these cost gains to benefit the auto industry and ultimately consumers.

The newly adopted business model by major EV makers to lease the battery separately from the vehicle is surely a necessary yet evolutionary step to considerably reduce the TCO. In order to reduce EV production costs the following requirements should be observed:

- Creation of a standard type of battery cell which would increase manufacturing volumes and lower costs.
- Increase EV range by optimising aerodynamics, kerb weight, tyre rolling resistance and energy management.
- Redeploy used EV batteries for stationary application using light commercial load, residential load and for distributed generation technologies such as renewables, wind and solar.
- Without redeployment and V2G integration, partially or fully electrified powertrains are still at a significant cost disadvantage over the entire lifecycle compared to conventional powertrains, in terms of the total cost of ownership (TCO) if seen from both supply and demand perspectives. A revised public policy and regulatory landscape may be necessary to shift this, and there needs to be encouragement of battery-related research and investment.

OEMs currently experience a shortfall in profit margins if they sell a PHEV rather than a vehicle with a conventional powertrain. Customers benefit from lower energy costs due to lower fuel consumption, but OEMs are not fully recompensed for the extra cost it incurs.

KPMG's 2013 Global Automotive Executive Survey hence suggests that 92 % believe that consumers' number one criteria will be fuel efficiency, with 36 % believing that plug-in hybrids will attract most consumer demand. 85 % of surveyed executives think that downsizing ICE engines is the solution, with a higher proportion investing there, though with a sizeable chunk of OEMs investing in plug-in hybrids, yet with only 8 % seeing battery technology as their biggest investment area [83]; something that needs to be seen in context of the advantages and disadvantages of different battery technologies [84, 85]. Energy efficient charging regimes of EVs will be important to electricity as fuel is to energy consumption [86].

9 Conclusions

According to the International Energy Agency (IEA), policy initiatives in 12 out of the 15 countries which are part of the Clean Energy Ministerial Electric Vehicle Initiative (EVI), have been put in place to boost the introduction of sustainable transport through a range of EV financial support measures and other practical facilitations [63] to stimulate this market.

There are a number of key major conditions to be met to increase the uptake of PHEVs and EVs:

- A significant reduction in cost of Lithium-ion battery and an increase in power density to provide EVs and PHEVs with an increased range. The IEA estimates that targeting a battery at $300 per kWh in 2020 [63] will make it competitive against an ICE.
- EVs and PHEVs price reduction through cheaper batteries and improved manufacturing processes will make EVs/PHEVs more competitive against ICEV. The IEA EV/PHEV Roadmap predicts that after 2015 the number of EVs/PHEVs will reach 7 million per year by 2020 [68]. If such a forecast is fairly reliable, manufacturing cost savings through larger volumes of production may be realized making these types of LCVs more appealing to own or lease.
- A broader development of national charging infrastructures through the widespread installation of public, commercial and private charging points. EVI countries are planning to install, as cumulative targets, about 2.4 million slow chargers and 6,000 fast chargers [63] by 2020.
- New business models applied to the use of EVs/PHEVs to lower customer's up front and monthly cost and to increase the availability of high power charging points in public and commercial environments. There is currently much uncertainty over the economics of rolling out and maintaining fast-charging infrastructure as investment is hardly profitable at low EV adoption rates, unless investment cost can be significantly lowered. Besides competition with alternative charging solutions (home and work), the general EV adoption rate is identified as being the main risk factor for private investment in public charging infrastructure. If private investment takes place at this premature stage, it appears to be driven by factors other than project prospects: Charging stations may be used as a perk to attract consumers with main revenue generated from non-electricity sales, such as commodity sales or to a certain extent parking fees. Integrated organizational structures with electric utilities promise slight improvements in return on investment since additional profits on the electricity market side enter the investment calculus. These additional profits are, however, very low. Fleet operation and grid tariff exemption can significantly improve returns [87].
- Intelligent charging for different profiles of users, and perhaps even using day-ahead management systems instead of pre-set profiles have desirable consequences for the system (e.g. decrease in variable costs, reduction in carbon emissions, increase of reliability) for the grid system [88], it is therefore necessary

to develop an "intelligent" charging strategy. Using an operation planning model, a study analyses the Spanish power system for 2020 under different EV penetration levels and charging strategies. The results show the benefits of using smart charging profiles instead of an unregulated profile, obtaining large cost reductions and maintaining system reliability levels [89].

Despite the technical and financial constraints for EV/PHEV adoption [90], it is worth noting that the latest EVs provide energy efficiency beyond 80 %, as compared to ICEV ($\sim$30 %) [66]. It remains clear that the toughest challenge to the large scale uptake of EVs/PHEVs in the forthcoming years is represented by the development of battery technology which can literally accelerate or stifle this evolution, with trade-offs between different battery technologies of the Lithium-ion family of battery technologies in terms of advantages and disadvantages related to safety, performance, specific energy, specific power, cost and lifespan. The second-life span and use of these batteries will also be of significance and consequence [91].

References

1. CE Delft (2011, April) Report, impacts of electric vehicles—deliverable 2, assessment of electric vehicle and battery technology. http://www.cedelft.eu/publicatie/impact_of_electric_vehicles/1153. Accessed 15 Nov 2013
2. Green Car Reports (2013, June 17) New gas mileage rules will reshape what Americans drive: aerodynamics and weight. http://www.greencarreports.com/news/1084852_new-gas-mileage-rules-will-reshape-what-americans-drive-aerodynamics-and-weight. Accessed 15 Nov 2013
3. Michalek J, Karabasoglu O (2013) Influence of driving patterns on lifecycle cost and emissions of hybrid and plug-in electric vehicle powertrains. Energy Policy 60:445–461
4. Energy Expert (2013, November 29) Electric cars not the answer now. http://www.jpost.com/Enviro-Tech/Energy-expert-Electric-cars-not-the-answer-now. Accessed 15 Nov 2013
5. Newsroom (2012, September 24) Toyota IQ EV. http://newsroom.toyota.eu/pressrelease/2941/1105/iq-ev. Accessed 15 Nov 2013
6. Fiat 500e (2013, April 16) http://www.topspeed.com/cars/fiat/2014-fiat-500e-ar138057.html. Accessed 16 Nov 2013
7. Fiat 500e is Nothing Short of Electrifying (2013, April 17). http://newcarpicks.com/site/?p=17152. Accessed 16 Nov 2013
8. Mercedes-Benz News (2011, December 28) Technology explained: blue efficiency. http://news.mercedes-benz.co.uk/innovations/technology-explained-blueefficiency.html. Accessed 16 Nov 2013
9. DRIVe For Business (2013) http://www.volvocars.com/uk/sales-services/sales/business-sales/drive-for-business/pages/default.aspx. Accessed 16 Nov 2013
10. Drag Coefficient (2013) http://www.calibrawiki.com/index.php?title=Drag_coefficient. Accessed 16 Nov 2013
11. Vehicle Coefficient of Drag List (2013) http://ecomodder.com/wiki/index.php/Vehicle_Coefficient_of_Drag_List. Accessed 15 Nov 2013
12. Scion iQ: is 37 MPG good enough? (2013) http://www.greencarreports.com/news/1063512_2012-scion-iq-is-37-mpg-good-enough. Accessed 18 Nov 2013
13. Scion iQ EV Specifications (2013) http://www.conceptcarz.com/vehicle/default.aspx?carID=22825&i=2. Accessed 18 Nov 2013

14. Nissan Versa Note (2014) http://www.cartype.com/pages/6206/nissan_versa_note__2014. Accessed 18 Nov 2013
15. BMW i3 (2013, July 30) The best city car on the market. http://www.plugincars.com/bmw-i3-may-be-best-city-car-market-127857.html. Accessed 18 Nov 2013
16. Mercedes-Benz B-Class Aerodynamics Tested with Paint-Video (2013) http://www.emercedesbenz.com/autos/mercedes-benz/b-class/mercedes-benz-b-class-aerodynamics-tested-with-paint-video/. Accessed 18 Nov 2013
17. 2012 Nissan LEAF—styling Review (2012, January 7) http://www.thecarconnection.com/review/1068274_2012-nissan-leaf_styling_2. Accessed 18 Nov 2013
18. AutoLabo Creates Aero Package for Toyota Prius (2008, September 12) http://jalopnik.com/5104713/autolabo-creates-aero-package-for-toyota-prius. Accessed 19 Nov 2013
19. Aero Prius Yurastyle (2013) http://priuschat.com/threads/aero-prius-yurastyle.37756/page-2. Accessed 19 Nov 2013
20. Ultimate Aero Mod (2013) Aero Prius Yura Style Neo kit. http://ecomodder.com/forum/showthread.php/ultimate-aero-mod-aero-prius-yura-style-neo-26709.html. Accessed 19 Nov 2013
21. The Toyota Prius has an Ugly Sister with a Sick Body! (2009, May 13) http://www.certifiedtoyotahybrids.com/the-toyota-prius-has-an-ugly-sister-with-a-sick-body/. Accessed 19 Nov 2013
22. The Toyota Prius gets HIT with the UGLY Stick! (2009, May 6) http://getbettermpg.blogspot.it. Accessed 21 Nov 2013
23. Advantages and Disadvantages of EVs (2013) http://auto.howstuffworks.com/fuel-efficiency/vehicles/electric-car-battery3.htm. Accessed 22 Nov 2013
24. How the Tesla Roadster Works (2013) http://auto.howstuffworks.com/tesla-roadster2.htm. Accessed 22 Nov 2013
25. Chevrolet Volt (2013) http://en.wikipedia.org/wiki/Chevrolet_Volt Accessed 19 Nov 2013
26. The Future of Urban Mobility (2013) BMW i. http://www.bmw.com/com/en/insights/corporation/bmwi/concept.html#lifedrive. Accessed 19 Nov 2013
27. Press Release—a New Era Dawns (2013, September 18) BMW group begins series production of the BMW i3 electric car in Leipzig. https://www.press.bmwgroup.com/global/pressDetail.html;jsessionid=TCtrSzhKL7LHL8FHYXLGzbQNkbLfFc6LtQPcLqKHqtbSdH76fX1S!261417453?title=a-new-era-dawns-bmw-group-begins-series-production-of-the-bmw-i3-electric-car-in-leipzig-industrial&outputChannelId=6&id=T0145644EN&left_menu_item=node__4088. Accessed 24 Nov 2013
28. UNSW School of Physics Sydney, Australia, Newton's laws (2013) Background and limitations. http://www.animations.physics.unsw.edu.au//jw/Newton.htm. Accessed 24 Nov 2013
29. Drag (physics) (2013) http://en.wikipedia.org/wiki/Drag_(physics). Accessed 27 Nov 2013
30. van Essen H, Kampman B (2011) Impacts of electric vehicles—summary report. http://www.ecologic.eu/8271. Accessed 5 Dec 2013
31. Bcatteries are Putting the Brakes on Electric Car Take-up (2010, June 21) http://www.theguardian.com/environment/cif-green/2010/jun/21/batteries-electric-cars. Accessed 5 Dec 2013
32. When Is A Charging Station Not A Charger? (2013) http://evworld.com/article.cfm?storyid=1893. Accessed 7 Dec 2013
33. Electric Car Onboard Charging Stations (2013) http://auto.howstuffworks.com/electric-car-onboard-charging-stations.htm. Accessed 7 Dec 2013
34. Clean cities December 2010 Webinar (2010, December 15) Electric vehicle charging levels and requirements overview, Idaho National Laboratory. http://www1.eere.energy.gov/cleancities/toolbox/pdfs/ev_charging_requirements.pdf. Accessed 8 Dec 2013
35. How to Charge Your Nissan Leaf (2013) http://www.nissan.co.uk/GB/en/vehicle/electric-vehicles/leaf/charging-and-battery/charging-nissan-leaf-and-battery.html. Accessed 8 Dec 2013
36. Bohnsack R, Pinkse J, Kolk A (2014) Business models for sustainable technologies: exploring business model evolution in the case of electric vehicles. Res Policy 42(2):284–300

37. Theory/How Far Do We Drive?/5 (2013) EV energy consumption. http://www.solarjourneyusa.com/EVdistanceAnalysis5.php. Accessed 9 Dec 2013
38. Farrington R, Rugh J, NREL (2000) Impact of vehicle air-conditioning on fuel economy, tailpipe emissions, and electric vehicle range. http://www.nrel.gov/docs/fy00osti/28960.pdf. Accessed 8 Dec 2013
39. Autocar (2013) Nissan Leaf performance and engineering. http://www.autocar.co.uk/car-review/nissan/leaf/performance. Accessed 11 Dec 2013
40. Battle of the Headlights (2010, November 16) Halogen vs. Xenon vs. LED, Bogdan Popa. http://www.autoevolution.com/news/battle-of-the-headlights-halogen-vs-xenon-vs-led-26530.html. Accessed 11 Dec 2013
41. Model S, Specs (2013) http://www.teslamotors.com/models/specs. Accessed 12 Dec 2013
42. CNET Reviews (2013, February 6) Nissan Leaf adds energy-efficient Bose stereo for 2013. http://reviews.cnet.com/8301-13746_7-57568016-48/nissan-leaf-adds-energy-efficient-bose-stereo-for-2013/. Accessed 13 Dec 2013
43. Nissan LEAF Drops Major New Beats with BOSE Energy Efficient Series Sound (2013, February 6) http://nissannews.com/en-US/nissan/usa/releases/e341da74-7279-4ad7-b0a1-135fbe74ee6a. Accessed 13 Dec 2013
44. Battery Monitoring/Management Systems (BMS) (2013) http://www.ev-propulsion.com/bms.html. Accessed 13 Dec 2013
45. Battery Management System (BMS) (2013) http://www.elektromotus.lt/en/products/battery-management-system-bms. Accessed 12 Dec 2013
46. Energy Saving Trust (2013) Guide to smarter driving in an electric car. http://www.google.co.uk/url?sa=t&rct=j&q=&esrc=s&source=web&cd=5&ved=0CFAQFjAE&url=http%3A%2F%2Fwww.energysavingtrust.org.uk%2Fcontent%2Fdownload%2F3286%2F68902%2Fversion%2F1%2Ffile%2FGuide%2Bto%2BSmarter%2BDriving%2Bin%2Ban%2BElectric%2BCar.pdf&ei=lzjAUv3QOOjNygO2ooHwDw&usg=AFQjCNEEcZwnqqhjdbLf2wUI1fraQZvnDQ. Accessed 13 Dec 2013
47. APRS Solar PHEV (2013) http://www.aprs.org/APRS-SPHEV.html. Accessed 10 Dec 2013
48. Automobile Accessory Power (2013) http://en.wikipedia.org/wiki/Automobile_accessory_power. Accessed 15 Dec 2013
49. High Volt Age (2012, November 22) Remember the 1990s push for 42-volt electrical systems? The Idea is back—maybe for good. http://blogs.motortrend.com/high-volt-age-remember-the-1990s-push-for-42-volt-electrical-systems-the-idea-is-back-maybe-for-good-26893.html. Accessed 15 Dec 2013
50. How to Reduce EV Production Costs? (2013) EV battery tech USA. http://www.cars21.com/news/view/670. Accessed 15 Dec 2013
51. Jiang T, Putrus GA, Gao Z, Conti M, McDonald S, Lacey G Development of a decentralized smart-grid charge controller for electric vehicles (prepared for publication)
52. Putrus GA, Bentley E, Binns R, Jiang T, Johnston D (2013) Smart grids: energising the future. Int J Environ Stud 70(5):691–701
53. Ekanayake J, Liyanage K, Wu J, Yokoyama A, Jenkins N (2012) Smart grid: technology and applications. Wiley, New Jersey, 2012
54. Marra F, Traholt C, Larsen E, Qiuwei W (2010) Average behavior of battery-electric vehicles for distributed energy studies. In: Innovative smart grid technologies conference Europe (ISGT Europe), IEEE PES, pp 1–7
55. Arrhenius Equation (2013) http://en.wikipedia.org/wiki/Arrhenius_equation#cite_note-IUPAC-ArrEq-0. Accessed 15 Dec 2013
56. Majima M, Satoshi U, Yagasaki E, Koyama K, SInazawa S (2001) Development of long life lithium ion battery for power storage. J Power Sour 101(1):53–59
57. Valence technology, lithium iron magnesium phosphate (LiFeMgPO4) battery modules, online datasheet
58. Lunz B, Yan Z, Gerschler J, Sauer D (2011) Influence of plug-in hybrid electric vehicle charging strategies on charging and battery degradation costs'. Energy Policy 46:511–519

59. Benjamin S et al (2010) LIFEMIT. In: IEEE conference in vehicle power and propulsion (VPPC)
60. Xia L, Mareels I, Alpcan T, Brazil M, de Hoog J, Thomas DA (2013) A distributed electric vehicle charging management algorithm using only local measurements. http://www.nicta.com.au/pub?doc=7351&filename=nicta_publication_7351.pdf. Accessed 15 Dec 2013
61. Tianxiang J (2013) PhD thesis, development of a smart grid interface controller for dynamic energy management of electric vehicles. Accessed 16 Dec 2013
62. International Energy Agency (2013, April) Global EV outlook 2013. http://www.iea.org/publications/freepublications/publication/GlobalEVOutlook_2013.pdf. Accessed 18 Dec 2013
63. AeroVironment Licenses PNNL EV Smart Charger Controller Technology (2013, March 6) http://www.greencarcongress.com/2013/03/pnnl-20130306.html. Accessed 16 Dec 2013
64. Bayram S, Michailidis G, Papapanagiotou I, Devetsikiotis M (2013) Decentralized control of electric vehicles in a network of fast charging stations. http://www4.ncsu.edu/~isbayram/globecomm13.pdf. Accessed 16 Dec 2013
65. European Commission (2013) Road transport: reducing CO_2 emissions from vehicles. http://ec.europa.eu/clima/policies/transport/vehicles/index_en.htm. Accessed 18 Dec 2013
66. Schipper L, Ng WS, Gould G, Deakin E (2013, March) Carbon in motion 2050 for North America and latin America, final report. http://metrostudies.berkeley.edu/pubs/reports/Carbon%20in%20Motion-UC-Final.pdf. Accessed 18 Dec 2013
67. International Energy Agency (2011, June) Technology roadmap, electric and plug-in hybrid electric vehicles. http://www.iea.org/publications/freepublications/publication/EV_PHEV_Roadmap.pdf. Accessed 18 Dec 2013
68. Nissan Leaf Prices and Specs (2013) http://www.nissan.co.uk/GB/en/vehicle/electric-vehicles/leaf/prices-and-equipment/prices-and-specifications.html. Accessed 18 Dec 2013
69. Model S Design Studio (2013) http://www.teslamotors.com/en_GB/models/design. Accessed 18 Dec 2013
70. Meet the 2012 Car of the Year (2013) Volt, the electric car with extended range. https://www.chevrolet.co.uk/cars/volt/. Accessed 18 Dec 2013
71. V60 (2013) http://www.volvocars.com/uk/all-cars/volvo-v60/Pages/default.aspx?utm_source=google&utm_medium=cpc&utm_term=buy%20v60&utm_campaign=V60%20-%20Brand%20-%20Generic%20-%20Broad. Accessed 18 Dec 2013
72. Toyota (2013) Price list December 2013, Prius. http://s3-eu-west-1.amazonaws.com/cdnlive.toyotaretail.co.uk/PriceList/Current_NewCarPrices_Dec13.pdf. Accessed 19 Dec 2013
73. Volvo V60 Plug-In (2012, May 14) http://www.autoexpress.co.uk/volvo/v60/18721/volvo-v60-plug. Accessed 18 Dec 2013
74. Update x3 (2013, May 29) This is HUGE! Chevy Spark EV could* cost as little as $12,495!. http://www.dailykos.com/story/2013/05/29/1212353/-This-is-HUGE-Chevy-Spark-EV-to-cost-as-little-as-12-495. Accessed 16 Dec 2013
75. Berger R (2009, September) Powertrain 2020—the future drives electric. http://www.zvei-elektromobilitaet.de/general/ecartec/2009-10-13_Urbschat.pdf. Accessed 18 Dec 2013
76. Chevrolet Volt Assembly Plant Begins Construction Of Solar Panel Arrays (2011, May 12) http://gm-volt.com/2011/05/12/chevrolet-volt-assembly-plant-begins-construction-of-six-acre-solar-panel-field/. Accessed 16 Dec 2013
77. At BMW (2013) Glue guns cut cost of new electric cars. http://online.wsj.com/news/articles/SB40001424127887323981304579081564191254956. Accessed 16 Dec 2013
78. Exceeding Expectation (2013) All the latest information and insights into the Mitsubishi Outlander PHEV. http://www.mitsubushi-cars.co.uk/newsletters/outlander-phev-201308.aspx. Accessed 16 Dec 2013
79. International Energy Agency (2011, June) Technology roadmap, electric and plug-in hybrid electric vehicles. http://www.iea.org/publications/freepublications/publication/EV_PHEV_Roadmap.pdf. Accessed 18 Dec 2013
80. Union of Concerned Scientists (2012) State of charge: electric vehicles' global warming emissions and fuel-cost savings across the United States. San Francisco, CA, 2012

81. Calnan P, Deane JP, Ó Gallachóir BP (2013) Modelling the impact of EVs on electricity generation, costs and CO_2 emissions, assessing the impact of different charging regimes and future generation profiles for Ireland in 2025. Energy Policy 61:230–237
82. Pasaoglu G, Honselaar M, Thiel C (2012) Potential vehicle fleet CO_2 reductions and cost implications for various vehicle technology deployment scenarios in Europe. Energy Policy 40:404–421
83. KPMG International (2013) KPMG's global automotive executive survey 2013. Managing a multidimensional business model. KPMG International, Switzerland
84. Catenacci M, Verdolini E, Bosetti V, Fiorese G (2013) Going electric: expert survey on the future of battery technologies for electric vehicles. Energy Policy 61:403–413
85. Salmien J, Kallio T, Omar N, van den Bossche P, van Mierlo J, Gualous H (2014) Transport energy—lithium ion batteries. In: Letcher TM (ed) Future energy. Improved, sustainable and clean options for our planet. Elsevier, Amsterdam, pp 291–309
86. Zhang L, Brown T, Samuelsen GS (2011) Fuel reduction and electricity consumption impact of different charging scenarios for plug-in hybrid electric vehicles. J Power Sour 196 (15):6559–6566
87. Rezania R, Pruggler W (2012) Business models for the integration of electric vehicles into the Austrian energy system. In: Proceedings of the 9th international conference on the European energy market (EEM), pp 1–8
88. Galus MG, Zima M, Andersson G (2010) On integration of plug-in hybrid electric vehicles into existing power system structures. Energy Policy 38:6736–6745
89. Manzie C (2009) Relative fuel economy potential of intelligent, hybrid and intelligent-hybrid passenger vehicles. In: Pistoia G (ed) Electric and hybrid vehicles. Power sources, models, sustainability, infrastructure and the market. Elsevier, Amsterdam, pp 61–90
90. Kotter R (2013) The developing landscape of electric vehicles and smart grids: a smart future? Int J Environ Stud 70(5):719–732
91. Lih W-C, Yen J-H, Shieh F-H, Liao Y-M (2012) Second-use applications of lithium-ion batteries retired from electric vehicles: challenges, repurposing process, cost analysis and optimal business model. Int J Adv Comput Technol 4:247–518

Part IV
Fleets

Evolution of E-Mobility in Carsharing Business Models

Susan A. Shaheen and Nelson D. Chan

Abstract Carsharing continues to grow worldwide as a powerful strategy to provide an alternative to solo driving. The viability of electric vehicles, or EVs, has been examined in various carsharing business models. Moreover, new technologies have given rise to electromobility, or e-mobility, systems. This paper discusses the evolution of e-mobility in carsharing business models and the challenges and opportunities that EVs present to carsharing operators around the world. Operators are now anticipating increased EV proliferation into vehicle fleets over the next 5–10 years as technology, infrastructure, and public policy shift toward support of e-mobility systems. Thus, research is still needed to quantify impacts of EVs in changing travel behavior toward more sustainable transport.

Keywords E-mobility · Carsharing · Electric vehicles · EVs · Business models · Station cars · Roundtrip carsharing · One-way carsharing

1 Introduction

Carsharing enables a group of members to share a vehicle fleet that is maintained, managed, and insured by a third-party organization. Primarily used for short-term trips, carsharing can provide affordable, self-service vehicle access 24-h per day for those who do not have a car, want to reduce the number of vehicles in their household, or do not use their vehicle during the day for long periods of time.

S.A. Shaheen (✉)
Transportation Sustainability Research Center, University of California, Berkeley, 408 McLaughlin Hall, Berkeley, CA 94720, USA
e-mail: sshaheen@berkeley.edu

N.D. Chan
Transportation Sustainability Research Center, University of California, Berkeley, 1301 South 46th Street, Building 190, Richmond, CA 94804-4648, USA
e-mail: ndchan@berkeley.edu

D. Beeton and G. Meyer (eds.), *Electric Vehicle Business Models*,
Lecture Notes in Mobility, DOI 10.1007/978-3-319-12244-1_10

Rates include fuel, insurance, and maintenance. Ideally, carsharing works best in a neighborhood, business, or campus setting where users could walk, bike, share rides, or take public transit to access the shared-use vehicles. Carsharing has evolved through several phases since the first carsharing system began in Europe in 1948. As of October 2012, carsharing was operating in 27 countries on five continents, with almost 1.79 million carsharing members sharing over 43,500 carsharing vehicles worldwide [1].

Recently, new business models have emerged due to increasing visibility and roundtrip carsharing usage (i.e., short-term auto access used for roundtrips), as well as new technologies and the development of electromobility, better known as e-mobility. Evolving aspects of e-mobility in carsharing include: electric vehicles (EVs) and scooters, gasoline-electric hybrid vehicles, keyless access, open-ended reservations, and the ability to conduct one-way trips. These have been catalysts to new e-mobility carsharing business models, which have increased membership and given carsharing greater flexibility and lower vehicle emissions [2]. EVs in particular appear to flourish in carsharing's pay-per-trip business models, distributing the high capital cost of the vehicle's electric battery from one driver to many. This chapter provides an overview of the evolution of e-mobility in carsharing, with a focus on developments in EV pilot programmes, academic research, and vehicle technologies. It also provides a framework for emerging business models and discusses model challenges and opportunities, particularly as e-mobility in carsharing continues to evolve.

2 Electric Vehicles in Carsharing

The e-mobility landscape consists of several business models to serve the needs of diverse members and land uses in the built environment. Each model presents opportunities and challenges. In this section, we discuss early station car programmes (i.e., vehicles primarily employed to provide a critical linkage between home, rail transit, and work locations). Next, EVs in roundtrip carsharing are explored. Several case studies of hybrid station car/roundtrip carsharing programmes are examined. The growth of EVs in one-way (also known as point-to-point or free floating) carsharing also is discussed. Finally, we provide lessons learned.

2.1 Station Car Programmes

EVs were a major part of station car programmes in the 1990s, particularly in the United States (U.S.) as a way to relieve parking constraints at rail transit stations [3]. The station car model placed shared vehicles at public transit stations, enabling its users to take transit to the station, and complete their trip with a station car, thus relieving the first-and-last mile problem. EVs were seen as an enhancement to the environmental benefits of station car programmes, since most trips were short distances suitable for zero-emission, all-electric propulsion.

The first large-scale station car program, Praxitèle, was launched in 1997 as an experimental demonstration in Saint-Quentin-en-Yvelines, a suburb of Paris. Overall, the EVs were well received by Praxitèle members who stated that the vehicles were compatible with the types of trips that they made [4]. Although the demonstration succeeded in its technical implementation, it struggled with costs and sustaining demand and ended after two years [5].

2.2 Roundtrip Carsharing

EVs also were integral to many roundtrip carsharing fleets in the 1990s [5]. Liselec launched in 1993 in La Rochelle, France to test EV use in carsharing. The program was successful and still exists today as Yelómobile, the longest operating EV carsharing program [6]. Yelómobile also operates as one-way carsharing, allowing users to drop off an EV at any of the program's charging locations rather than returning it to its original station. Since trips stay within La Rochelle, they keep within the EV maximum range of 130 km (81 miles). Unlike other EV carsharing programmes struggling with economic sustainability, Yelómobile continues to receive governmental support for its operations [7].

Many carsharing programmes in Asia, particularly Japan, began as project-based, EV carsharing programmes to provide mobility to downtown business customers [8]. The Second Car System (SCS) launched in the Tama New Town District of Inagi City in Japan in 1999. Most of its 300 users reported that the service met their travel needs. SCS included a reservation system that calculated the charge time needed to complete a planned trip and verified that a vehicle with enough battery life was available before confirming a reservation. In this way, the limited vehicle range of 160 km (99 miles) did not present an obstacle. However, the program lost users after implementing fees and closed after three years [9].

Automakers have experimented with EV carsharing as well. Toyota Motor Company deployed the Crayon System in Toyoda City, Japan in the late 1990s. Toyota's employees used the program's ECom vehicles for business trips or for commuting between home and the office. The program consisted of 50 vehicles, 13 stations, and 700 members [10]. Crayon employed advanced ITS technologies including: (1) automated reservations, (2) telematics to communicate between the vehicles and system management, and (3) GPS technology to track the cars. Nissan Motor Company also entered into EV carsharing in 1997, with the Minato-Mirai 21 experiment in Yokohama. The program's field studies began in 1999 and grew to 30 vehicles and seven stations in the Yokohama area. The trials ended in March 2002, and the system transitioned to operators to determine system viability. This program spread to Yokohama, Kawasaki, and Tokyo and was called the Intelligent Transportation System/Carsharing Electric Vehicle (ITS/CEV) City Car System [11, 12].

It later became known as OrixCarsharing, comprising from 6,000 to 8,000 members sharing approximately 400 vehicles. OrixCarsharing discontinued exclusive EV use and now employs gasoline vehicles and gasoline-electric hybrids [13].

Since these early programmes, EV carsharing has waned in Asia, with the industry evolving toward the roundtrip carsharing model. Moreover, it has drastically reduced its use of EVs in carsharing fleets. Outside of Japan, Singapore, and South Korea, carsharing has experienced slower growth than in Europe and the Americas [8].

2.3 Hybrid Station Car/Carsharing Models

Several research studies investigated the viability of blending the concepts of station cars and carsharing to create a hybrid model. One was the CarLink field test, which ran from 1999 to 2000, which deployed 12 natural gas Honda Civics at the Dublin-Pleasanton Bay Area Rapid Transit (BART) District station in the San Francisco Bay Area. The program was deemed a success from an operational and user perspective [14, 15]. CarLink II followed this demonstration and was deployed from 2001 to 2002 at the Caltrain station in Palo Alto, California [16]. Flexcar (later merged with Zipcar in 2007) took over CarLink in 2002, but closed the service in 2003 due to financial concerns. Another BART station car initiative was the Hertz station car program, which included two Th!nk City EVs at the Fremont BART station from 2000 to 2003 [3].

Several research programmes at the University of California (UC) have also piloted hybrid models. UC Irvine continues to run the Zero-Emission Vehicle Network-Enabled Transport (ZEV•NET) research program today. Deployed in 2001, ZEV•NET enhances mobility from the Irvine Transportation Center commuter rail terminal to the employment sites of four companies and UC Irvine. The current fleet is comprised of Toyota RAV4 EVs, Mitsubishi iMiEVs, and Scion iQ EVs [17]. In addition, UC Riverside deployed an EV carsharing pilot named Intellishare, which began in 1999 and ended in July 2010 [18]. Intellishare was similar to ZEV•NET, as it added a station element to its EV carsharing system. Station cars were located at the downtown Riverside Metrolink train station and could be reserved for transport to the UC Riverside campus [19].

2.4 One-Way Carsharing

One-way carsharing—where members are not required to return a shared-use vehicle to the same station from which they borrowed it—began in Europe in the 1970s as experiments, with Procotip in Montpellier, France and Witkar in Amsterdam. Due to the lack of technology for system rebalancing and limited governmental support, these experiments failed after several years. One-way services resurfaced in the late-1990s under the station car model (see Sect. 2.1). The Praxitèle demonstration in Paris found that 90 % of trips were one-way trips [20].

In the U.S., UC Riverside's Intellishare program (mentioned earlier) was deployed as a one-way EV carsharing pilot. Intellishare's fleet had high usage—averaging 100 daily trips. To ensure a sufficient charge on the EVs (i.e., the EVs had a range of 160 km or 99 miles), the system would not allow depleted vehicles to be available for use until they finished charging. In this way, the limited EV range was not problematic; nevertheless, the project ceased operations in 2010 and was not commercialized [21].

Recent growth in modern one-way carsharing has been primarily in free-floating carsharing and one-way rentals between airports and cities. Daimler started the first free-floating EV carsharing service in October 2008 in Ulm, Germany, known as car2go. Its success has enabled international expansion. Since 2010, car2go has expanded throughout Western Europe, the U.S., and Canada. Globally, car2go has a fleet of 7,300 gasoline vehicles and 1,000 Smart Fortwo EVs, with 375,000 members [22]. BMW-Sixt launched a free-floating carsharing system in 2011 in Munich known as DriveNow and has since expanded to Berlin, Düsseldorf, Cologne, and San Francisco. Approximately 1,000 BMW ActiveE EVs and gasoline vehicles are accessible by 60,000 members. Both car2go and DriveNow have worked with cities to prepay for parking spaces for their free-floating vehicles. As of July 2013, one-way carsharing represented 12 and 16 % of North American carsharing membership and fleets deployed (roundtrip carsharing and one-way aggregate totals that do not include peer-to-peer carsharing), respectively [23]. In December 2011, Autolib' was launched in Paris, France by Bolloré. Today, Autolib' has almost 30,000 members accessing 1,800 Bluecar EVs at 800 stations throughout the Paris metropolitan area. The system boasts over two million trips taken [22]. Most recently, Communauto launched Auto-mobile, a pilot project planned from June to October 2013. The project consisted of a fleet of 20 EVs shared in a neighborhood of Montréal, Canada. In October 2013, Auto-mobile expanded to another neighborhood with plans for a third [24].

At the time of this writing, there approximately 11 one-way carsharing operators worldwide, with programmes in Austria, Canada, France, Germany, Japan, Mexico, Spain, the United Kingdom (UK), and the U.S. Several more systems are planned for launch in the next several years, notably, one in China.

2.5 Lessons Learned

Although most e-mobility programmes proved to be feasible in terms of driving range and user satisfaction, EVs gradually faded out of station car, roundtrip carsharing, and hybrid systems. Although EVs were noted as a successful part of station car systems, 60 % of all programmes ceased in the early 2000s [25]. By 2006, the vast majority of EVs in carsharing programmes had disappeared in favour of gasoline-electric hybrid vehicles. Numerous reasons were catalogued for failure:

high costs; high insurance rates; low reliability of the first generation EVs; a preference for hybrid vehicles; decreased user demand and public support; operational barriers (e.g., limited vehicle range, few charging stations); logistical challenges (i.e., the need for centralized management and real-time data feedback); and economic downturn [13]. Nevertheless, shared-use mobility services have experienced a recent resurgence in EVs. Due to technological advancements, automakers have launched next-generation EVs at lower costs with longer-range batteries, such as lithium-ion.

Another key understanding involves the role of public policy in acting as a major catalyst in the introduction of EVs into vehicle fleets. In California, the Zero Emission Vehicle (ZEV) program was designed to achieve the state's emission reduction goals and requires automakers to sell more ZEVs (i.e., vehicles with zero tailpipe emissions, including EVs). Automakers can also receive additional ZEV sales credit by placing them in transportation systems that demonstrate technology-enabled vehicle sharing, such as carsharing programmes [26]. Moreover, monetary rebates encourage carsharing operators to purchase ZEVs and other low-emission vehicles [27]. Supportive parking policies also play a role in supporting e-mobility. Cities in Australia, North America, and several European countries have provided free or discounted on-street parking to carsharing operators as a form of non-monetary support [28]. Agencies such as the Port of San Francisco have also considered mandating EV charging station allocations in lease renewals of off-street parking [29]. Finally, cities can consider investing in electric charging infrastructure to encourage EV proliferation in carsharing.

3 Current and Projected Growth of EV Carsharing

Building upon lessons learned from previous generations of EVs and carsharing models, many worldwide carsharing experts believe a trend over the next five years will be the re-emergence and growth of EVs in e-mobility fleets. Worldwide surveys of carsharing operators conducted in 2006, 2008, and 2010 noted a shift in vehicle propulsion between 2006 and 2010 toward gasoline-electric hybrid vehicles and EVs [30]. Several key trends are occurring: (1) automakers are taking a lead in launching e-mobility systems in cities in Europe, Japan, and North America and (2) existing carsharing operations are reintroducing EVs into their fleets.

3.1 E-Mobility Systems by Automakers

Automaker-sponsored e-mobility systems currently are a significant portion of carsharing membership and fleets. In January 2013, Daimler's car2go and BMW-Sixt's DriveNow represented 11.7 and 18.4 % of North American carsharing membership and fleets deployed, respectively [31]. Automakers are continuing to integrate EVs into new and existing carsharing operations. Nissan and the City of Yokohama launched Japan's first one-way carsharing program in October 2013, with 30 Nissan New Mobility CONCEPT EVs [32]. Renault began Twizy Way in 2012, a carsharing pilot in Saint-Quentin-en-Yvelines near Paris employing 50 Twizy EVs [33]. In September 2013, Renault transferred Twizy Way to web services and an automotive engineering company, Keymoov, to operate [34]. Also in September 2013, Renault and Bolloré began a joint venture to launch commercial and industrial EV carsharing programmes in France and abroad. Moreover, they plan to develop a three-seater EV with a 200-km (124-miles) range, after Autolib' data found that 75 % of rentals involve three passengers or less [35]. Daimler continues to operate car2go in Europe and North America, with plans for continued expansion, many of which include EVs in their fleets. Peugeot Citroën Automobiles launched Multicity in Berlin—the first all-EV carsharing system in Germany—in September 2013 with 350 C-Zero EVs [36]. Bolloré, Daimler, Citroën, and BMW-Sixt are the top four EV carsharing operators, with a total of 3,280 EVs [22].

3.2 Re-Emergence into Existing Carsharing Fleets

Following the trends noted in the biannual worldwide carsharing survey [30], EVs have begun to re-emerge into existing carsharing fleets. Carsharing operators have been adding EVs into their programmes since 2011. Hertz 24/7™ (formerly Hertz On Demand) deployed Mitsubishi i-MiEVs in the UK in April 2011 [37]. Similarly, City CarShare deployed i-MiEVs in the San Francisco Bay Area in December 2011 [38] and maintain approximately 60 EVs in their fleet. Other recent EV carsharing introductions throughout North America include: Communauto's Auto-mobile in Montréal, I-GO CarSharing (now part of Enterprise CarShare) in Chicago, and Zipcar (now part of Avis Budget Group). Most recently, City CarShare in partnership with Toyota and the Transportation Sustainability Research Center at the University of California, Berkeley began a three-year EV carsharing pilot called Dash™ in September 2013. Scion iQ EVs have been placed at the Hacienda Business Park in Pleasanton, California, for employees and residents to use for short-distance trips [39].

Carsharing programmes outside of North America are continuing to expand into new cities, as well. After the success of Autolib' in Paris, Bolloré launched a similar program in Lyon in October 2013, and plans another EV carsharing program in Indianapolis, Indiana in late-2014 [22].

One country to note the emergence of EVs carsharing in is Australia, where EVs have had low sales [40]. In 2012, GoGet CarShare, the largest carsharing program in Australia, deployed Mitsubishi i-MiEVs and Nissan Leafs® into their Melbourne fleet [41].

As of late-2013, EVs have been deployed in carsharing programmes in approximately 14 countries: Australia, Austria, Denmark, Finland, France, Germany, Italy, Japan, Norway, the Netherlands, Portugal, Switzerland, the UK, and the U.S.

4 Conclusion

Carsharing operators continue to anticipate greater EV potential in e-mobility systems in the future. At present, EVs have been introduced into over half of the 27 countries where carsharing currently operates. E-mobility operators can employ next-generation EVs with longer travel range, but they must maintain sufficient charge across the vehicle fleet to support successful operations. To address this, some operators are deploying EVs in one-way programmes in higher-density areas to support shorter usage and rebalancing trips. Moreover, many operators are working with municipal governments to influence public policy to gain access to public transit and on-street station parking, as well as EV charging infrastructure. Governmental support can play a notable role in encouraging e-mobility systems and helping to achieve greenhouse gas emission reduction targets through policy adaptation, public outreach, and financial assistance. In light of growing EV demand, charging infrastructure, technological advance, supportive policies, and shared-use mobility integration, e-mobility systems are poised to impact travel behavior in many regions across the globe. Research is needed to quantify and understand e-mobility impacts on vehicle kilometers/miles traveled, household vehicle holdings, and modal shift and to guide future policymaking.

References

1. Shaheen S, Cohen A (2012) Innovative mobility carsharing outlook: carsharing market overview, analysis, and trends: fall 2012. Transportation Sustainability Research Center, Institute of Transportation Studies, University of California, Berkeley
2. Schwieger B, Wagner C (2012) Second generation car-sharing: developing a new mobility services target groups and service characteristics. Südwestdeutscher Verlag für Hochschulschriften AG Company KG
3. Barth M, Shaheen S (2001) Shared-use vehicle systems: a framework for classifying carsharing, station cars, and combined approaches. Transp Res Rec J Transp Res Board 1791:105–112
4. Massot M, Allouche J, Bénéjam E, Parent M (1999) Praxitèle: preliminary results from the Saint-Quentin station-car experiment. Transp Res Rec J Transp Res Board 1666:125–132
5. Shaheen S, Sperling D, Wagner C (1999) A short history of carsharing in the 90's. J World Transp Policy Pract 5(3):18–40
6. Communauté d'Agglomération de La Rochelle (2012) Yélo. http://yelo.agglo-larochelle.fr/accueil. Accessed 23 Oct 2014
7. McDonald M, Vöge T (2001) Analysis of potentials and limitations of cybernetic transport systems—combined deliverable for D1.1—system operating scenarios and D1.3—barriers to deployment. Technical report, cybermove—cybernetic transportation systems for the cities of tomorrow
8. Wang M, Martin E, Shaheen S (2012) Carsharing in Shanghai, China: analysis of behavioral response to local survey and potential competition. Transp Res Rec J Transp Res Board 2139:86–95

9. Fukuda T, Kashima S, Barth M (2003) Evaluating second car system, an electric vehicle sharing experiment in Tama New Town District, Inagi city, Tokyo. In: Presented at the TRB annual meeting
10. Barth M (2001) Shared-use station car programs. In: Presentation at shared-use, station car summit from University of California, Irvine
11. Takayama M (2002) Transportation and safety in Japan: introduction of the ITS/EV city car system. J IATSS Res 26(2):118–121
12. Barth M, Shaheen S, Fukuda T, Fukuda A (2007) Carsharing and station cars in Asia: an overview of Japan and Singapore. Transp Res Rec J Transp Res Board 1986:106–115
13. Cohen A, Shaheen S, Brook D, Martin E (2008) Light electric vehicles (LEV) project: consulting final report. Industrial Technology Research Institute
14. Shaheen S (1999) Dynamics in behavioral adaptation to a transportation innovation: a case study of CarLink—a smart carsharing system. Institute of Transportation Studies, University of California, Davis, UCD-ITS-RR-99-16:232
15. Shaheen S, Wright J, Dick D, Novick L (2000) CarLink—a smart carsharing system field test final report. Institute of Transportation Studies, University of California, Davis Publication, UCD-ITS-RR-00-4:182
16. Shaheen S, Novick L (2005) Framework for testing innovative transportation solutions: case study of CarLink, a commuter carsharing program. Transp Res Rec J Transp Res Board 1927:149–157
17. ZEV•NET (2010) FAQ, University of California, Irvine. http://zevnet.fastfleet.net/how/faqs/#vehicles. Accessed 26 July 2012
18. South Coast Air Quality Management District (2010) AQMD to honor clean air heroes at 22nd annual clean air awards. South Coast Air Quality Management District
19. Shaheen S, Sperling D, Wagner C (1998) Carsharing in Europe and North America: past, present, and future. Transp Quart 52(3):35–52
20. Massot H (2000) Praxitèle, un concept, un service, uneexpérimentation, bilan d'un prototype. la revue française TEC, no 159, May–July 2000
21. Barth M, Todd M, Murakami H (2000) Intelligent transportation system technology in a shared electric vehicle program. In: Transp Res Rec J Transp Res Board 1731:88–95
22. Fairley P (2013) Car sharing could be the electric vehicle's killer app. IEEE Spectrum, 21 Aug 2013. http://spectrum.ieee.org/transportation/advanced-cars/car-sharing-could-be-the-electric-vehicles-killer-app. Accessed 25 Aug 2013
23. Shaheen S (2013) Carsharing snapshot and impacts: the Americas. In: Presentation at the shared-use mobility summit, San Francisco, California (Unpublished data)
24. Communauto (2013) Auto-mobile arrives in Rosemont and Côte-des-Neiges-Notre-Dame-de-Grâce. Communauto, Press Release, 17 Oct 2013. http://actualites.communauto.com/en/2013/10/17/auto-mobile-arrives-rosemont-cote-des-neiges-notre-dame-de-grace/. Accessed 23 Oct 2013
25. Shaheen S, Schwartz A, Wipyewski K (2004) Policy considerations for carsharing and station cars: monitoring growth, trends, and overall impacts. Transp Res Rec J Transp Res Board 1887:128–136
26. Shaheen S, Wright J, Sperling D (2002) California's zero emission vehicle mandate: linking clean-fuel cars, carsharing, and station car strategies. Transp Res Rec J Transp Res Board 3857:113–120
27. California Environmental Protection Agency Air Resources Board (2012) Advanced clean cars: 2012 proposed amendments to the California zero emission vehicle program regulations. California Environmental Protection Agency Air Resources Board
28. Shaheen S, Cohen A, Martin E (2010) Carsharing parking policy: a review of North American practices and San Francisco, California, Bay Area case study. Transp Res Rec J Transp Res Board 2187:146–156
29. Moyer M (2009) Parking bid resolution memorandum. City and county of San Francisco Port Commission, 5 Mar 2009. http://www.sfport.com/ftp/uploadedfiles/meetings/supporting/2009/Item%2010A%20Parking%20Lotspdf. Accessed 23 Oct 2014

30. Shaheen SA, Cohen AP (2013a) Carsharing and personal vehicle services: worldwide market developments and emerging trends. Int J Sustain Transp 7(1)5–34
31. Shaheen S, Cohen A (2013b) Innovative mobility carsharing outlook: carsharing market overview, analysis, and trends: summer 2013. Transportation Sustainability Research Center, Institute of Transportation Studies, University of California, Berkeley
32. Nissan Motor Company (2013) Nissan and Yokohama city launch Choimobi Yokohama, Japan's first one-way car sharing service with a large number of ultra-compact EVs. Press release, Yokohama, Japan. http://www.nissan-global.com/EN/NEWS/2013/_STORY/131010-02-e.html. Accessed 30 Oct 2013
33. Renault (2012) Twizy Way by Renault: an innovative shared electric mobility service. Press release, Renault Group, 11 June 2012. http://media.renault.com/global/en-gb/renault/Media/PressRelease.aspx?mediaid=32671. Accessed 1 Oct 2013
34. Renault (2013) Renault calls on Keymoov to operate the Twizy Way car-sharing service in Saint-Quentin en Yvelines, France. Press release, Renault Group, 24 Sept 2013. http://media.renault.com/global/en-gb/renaultgroup/Media/PressRelease.aspx?mediaid=50836. Accessed 1 Oct 2013
35. Mihalascu D (2013) Renault and Bolloré to develop a 3-seater electric car, EV-sharing services. CarScoops, 16 Sept 2013. http://www.carscoops.com/2013/09/renault-and-bollore-to-develop-3-seater.html. Accessed 15 Oct 2013
36. CITROËN (2013) Multicity Carsharing. CITROËN multicity carsharing. https://www.multicity-carsharing.de/en/. Accessed 15 Oct 2013
37. Hertz (2013) Hertz unveils 'Hertz On Demand' redefining car sharing. Press release, Park Ridge, NJ. http://www.hertz247.com/NewYork/en-US/About/Article/2?artid=408. Accessed 23 Oct 2014
38. City CarShare (2011) Mitsubishi motors makes first fleet delivery of the 2012 Mitsubishi i-MiEVs (Mitsubishi innovative electric vehicle) to Bay Area's City CarShare. San Francisco, Press Release, City CarShare, 8 Dec 2011. https://www.citycarshare.org/article/mitsubishi-first-fleet-delivery/. Accessed 21 Sept 2013
39. City CarShare (2013) Jetting out for a quick trip? Introducing DASH. https://www.citycarshare.org/dash/. Accessed 21 Sept 2013
40. Blackburn R (2012) Unplugged: electric cars losing their lustre. The Sydney morning herald, 18 Jan 2012. http://www.smh.com.au/drive/motor-news/unplugged-electric-cars-losing-their-lustre-20120118-1q5d6.html. Accessed 23 Oct 2014
41. Hopewell L (2013) GoGet has electric share cars now in a great leap forward for Aussie EVs. Gizmodo Australia, 19 July 2013. http://www.gizmodo.com.au/2013/07/goget-has-electric-share-cars-now-in-a-great-leap-forward-for-aussie-evs/. Accessed 29 Oct 2013

Personalized Total Cost of Ownership and Range-Capability Assessment as an EV Sales Accelerator

Sunny Trochaniak, Megan Allen, Eric Mallia, Jennifer Bauman and Matthew Stevens

Abstract Several activities are currently underway that will further reduce the Total Cost of Ownership (TCO) and improve the user experience for electric vehicles. While this will potentially increase the favorability of EVs, this will only translate to increased EV adoption if potential EV owners are informed of these benefits in a method that is understandable and personally relevant. Multiple studies have concluded that current methods are insufficient in providing this awareness and that the lack of suitable alternatives will delay EV adoption. This paper will present the findings of a new method of EV purchase assistance, highlighting key findings from over 20 fleets that have implemented the system.

Keywords Electric vehicles · Total cost of ownership · High efficiency vehicles · Fuel economy · Vehicle data logging

S. Trochaniak (✉)
MyCarma, 2-60 Northland Rd, Waterloo, ON N2V 2B8, Canada
e-mail: sunny@mycarma.com

M. Allen · E. Mallia
FleetCarma, 2-60 Northland Rd, Waterloo, ON N2V 2B8, Canada
e-mail: mallen@fleetcarma.com

E. Mallia
e-mail: emallia@fleetcarma.com

J. Bauman · M. Stevens
CrossChasm Technologies, 2-60 Northland Rd, Waterloo, ON N2V 2B8, Canada
e-mail: jbauman@crosschasm.com

M. Stevens
e-mail: mstevens@crosschasm.com

D. Beeton and G. Meyer (eds.), *Electric Vehicle Business Models*,
Lecture Notes in Mobility, DOI 10.1007/978-3-319-12244-1_11

1 Why Does Personalization Matter?

There are a number of initiatives that are currently underway that aim to further reduce the Total Cost of Ownership (TCO) and improve the user experience for electric vehicles (EVs). While this will increase the theoretical favorability of EVs, this will only translate to increased EV adoption if potential EV owners are informed of these benefits in a method that is understandable and personally relevant.

With higher efficiency vehicles and electric vehicles offered at a higher price point than their conventional counterparts, the purchaser is required to justify the additional expense. Without personalized results this justification becomes a task that the purchaser has to complete on their own, and based on assumptions that may not actually apply to them [1]. Previous studies have identified that car buyers put a high risk premium on the energy savings portion of the calculation and demonstrate highly risk-averse behaviour [2]. The result is lower adoption of high fuel efficiency vehicles than economically sensible. Given the additional risks associated with electric vehicles it is anticipated that the risk aversion tendency will be amplified.

At its core, the risk premium applied to fuel savings is rooted in two constraints of current fuel efficiency labels; the generic nature and unit confusion. Due to their primary intent, fuel efficiency labels are based on laboratory results of vehicles run on dynamometers on generic drive cycles. This approach provides a standardized test procedure that enables repeatable and auditable results. While there are strong benefits of this approach, there are two challenges in relation to use by individual vehicle purchasers. The first is that due to the number of stakeholders involved and the cost to develop and change test procedures, it is very difficult to add or modify the test cycles as operating patterns and technology changes. An example of this is the significant delay in the inclusion of air-conditioner use into the test cycles in North America. The second is that the inherent and necessary use of a generic cycle generates the same generic result for a driver in downtown New York City as for a driver in rural Alaska. As a consequence, many vehicle buyers have substantial scepticism towards the fuel efficiency label's ability to provide them a fuel efficiency number that will approximate their personal situation.

The second challenge is unit confusion. Substantial effort has been made by regulators to generate fuel efficiency labels that are as intuitive and standardized as possible. This task is becoming increasingly complex as powertrain technology evolves. For example, the EPA Monroney label for plug-in hybrids makes every attempt to assist a car buyer in understanding the potential efficiency. As a result, both the electric and gasoline consumption values are presented in mile-per-gallon-equivalent and miles-per-gallon (MPG) numbers respectively. While MPG is inherently understood, the use of that unit as an electric equivalent provides substantial confusion. To calculate energy costs the consumer must understand the electricity cost of a gallon-equivalent. While the label provides an annual fuel estimate, few car buyers can actually replicate that calculation from the MPG and MPGe results provided. Accordingly, car buyers require the fuel efficiency results to be presented in the most useful units, which will generally be energy cost per month or Total Cost of Ownership.

Utilizing individualized results instead of commonly available generic fuel efficiency results based on dynamometer testing has been presented as a solution to the risk aversion problem. Whether the buyer is purchasing a vehicle for personal use or a specific vehicle within a commercial fleet, personalization de-risks the case for electric vehicle purchase. When assumptions and not data are used to make a recommendation to purchase a more expensive higher efficiency or electric vehicle, it may lead to the purchaser feeling as if they are taking a greater risk. Without confidence that the vehicle will both suit their needs as well as reduce expenses, the choice to purchase an electric vehicle becomes more difficult.

The personalization of vehicle recommendations makes a more compelling case for adoption. Several stakeholders that recognize this strategy have stepped forward, including utilities, municipalities and environmental organizations. Utilities in particular can benefit from increased electric vehicle adoption as it opens a new market, and through personalized recommendations they are able to engage with users in their service area. Utilities can also learn about the ways in which electric vehicle sales work, common objections, and the strengths of implementation.

When an individual is informed of their personalized recommendation, it is more compelling to deliver the message in terms they understand. For individuals buying a vehicle for personal use, the way to communicate the value of an electric vehicle resonates most when put into terms of monthly fuel savings. Personal car buyers often pay financing, lease, and/or insurance payments on a monthly basis, and so these EV benefits and savings need to be conveyed in a similar dollars per month figure.

Combining an increased monthly payment with a reduced monthly fuel cost helps make definitive comparisons between a higher efficiency or electric vehicle and other vehicles they are considering. Demonstrating the value of an electric vehicle as a reduced overall monthly payment puts the recommendation and messaging in the exact terms that are meaningful to the individual purchaser.

For commercial fleets, the Total Cost of Ownership for a vehicle over its lifetime holds the greatest relevance. With a common life cycle of 7 years, it is important for fleets to consider all elements beyond just the base purchase price of a vehicle. This includes ownership costs such as incentives, resale value and depreciation, financing, insurance, administration, fuel and energy costs, and finally parts and service work. In some cases, fleets may also be taxed for the amount of greenhouse-gases they emit. All of these factors are especially important to be considered because they have a multiplying effect on the Total Cost of Ownership over the lifespan of the vehicle.

While some fleets may already have established processes for vehicle procurement based on Total Cost of Ownership, the calculations become much more complex when considering the purchase of higher efficiency or electric vehicles.

Incentives are often significant, and can vary greatly from region to region. Resale values are less predictable. The cost difference between the price of gasoline and energy and other technologies also becomes much more relevant. When comparing two conventional internal combustion engine vehicles, an increase in the price of gasoline may have a minimal impact, but when comparing a gasoline vehicle to an electric vehicle, this price increase directly impacts the bottom line.

With higher efficiency and electric vehicles possessing different components to their conventional counterparts, maintenance calculations become more dynamic. Electric vehicles eliminate the need for certain regularly scheduled maintenance as there are fewer fluids to change, and brake wear is significantly reduced due to regenerative braking, however they also introduce a new element with battery maintenance.

Range capability with electric vehicles introduces an entirely new variable to the Total Cost of Ownership. Electric vehicles vary with their proposed range capabilities, and this variance increases further when factoring in the effects of different driving behaviour, driving routes, and weather conditions. Their real-world ranges can severely differ from what was advertised based on those factors, making range capability much more personal.

With conventional vehicles, unless the vehicle is being driven through remote locations, running out of fuel simply means stopping at the nearest gas station 'around the corner' to fill up. With electric vehicles, running out of battery can hold much more severe consequences. With the current infrastructure providing limited support for charging stations, the cost of running out of battery is not only the cost of towing it to the nearest charging station, but also the opportunity cost of the driver not being able to complete their job.

Considering all of these factors, the question may not only be 'how much can I save with an electric vehicle?' but also 'what is the cost of making a bad decision?' Since there can be such a significant financial impact based on whether a good or bad decision is made, on a decision which has traditionally been made through assumptions and high uncertainty, a case for personalization is made. Personalization greatly reduces this uncertainty, and enables fleets and individuals to make sound financial decisions while reducing their adverse effects on the environment.

2 How Does It Work?

2.1 Introduction

2.1.1 Model-Based Design

Model-Based Design (MBD) is the core technical element of the personalization approach that is described in this paper. Model-Based Design was originally developed in the aerospace industry as a core development tool. The approach is based on the significant use of software models in the development of a design. By creating a physics-based, or cyber-physics based software model of rockets, the designs could be simulated, analyzed, and improved in a software environment. The use of software models reduces (though rarely eliminates) the need for prototype development and thereby drastically reduces development costs, development time, and increases confidence in the ultimate performance of the final design.

Over the past decade the practice of model-based design in the automotive industry has become widely adopted. The beachhead for this technology was hybrid and plug-in hybrid vehicles. Given the significant number of design variables in these powertrains there is a high need for tools and systems to evaluate the impact of the various design decisions. Accordingly, the traditional approach of significant prototyping was cost and time prohibitive. The use of MBD has now expanded to all types of powertrains given the cost, time, and quality benefits.

Traditional use of MBD has been to generate an optimal vehicle design. Specifically, it has been used to estimate the impact of powertrain design decisions on the resulting energy consumption and performance. This technology is equally capable of estimating the personalized energy consumption and performance of a given production vehicle on a personal drive cycle. This approach has been developed and tested by FleetCarma.

FleetCarma is a division of CrossChasm Technologies. Founded in 2007, CrossChasm provides engineering support to major OEMs in the vehicle design process and control system integration. A core component of CrossChasm's work for OEMs has been Model-Based Design. FleetCarma leveraged this expertise to create a service for fleets based on real-world data logging, prediction, and results.

FleetCarma currently focuses on two core service offerings. The first employs data loggers within a fleet to accurately measure real-world fuel consumption and characteristics of a vehicle's duty cycle. FleetCarma's predictive modelling and simulation accurately assess the performance of electric vehicles within the logged duty cycle. The second core service FleetCarma provides is in-service performance monitoring. FleetCarma data loggers access hard-to-get data on electric vehicles, such as battery state-of-charge and electricity consumption, and provide insights with that data to key stakeholders, customers, and partners.

2.1.2 Model Library Generation

In the MBD approach, to be able to estimate the personal energy consumption on a specific vehicle it is required that a software model of that production vehicle be created. To accomplish that, a production vehicle is obtained by the FleetCarma team and that vehicle is driven in real-world conditions over a specific set of tests. These tests have been designed to accomplish two goals: (a) provide the software modelling team the vehicle characteristics required to make an accurate vehicle model (i.e. control logic patterns), and (b) data for model validation.

This process originally took approximately 2 weeks of testing and 2 weeks of modelling time. Over multiple development iterations, this process is now down to less than a day for conventional, hybrid, and electric vehicles. Three days is required for plug-in hybrids.

As of writing this paper the model library included 176 vehicle models, spanning production battery-electric, plug-in hybrid, hybrid, and conventional vehicle models.

2.1.3 Electric Vehicle Suitability and Costing

Through expertise in powertrain design, FleetCarma developed a service in which vehicle models could simulate the performance of electric vehicles in the real world. This analysis not only predicts the fuel and power consumption of electric vehicles but predicts the range and charge capabilities of the simulated vehicles. This analysis involves running the real-world drive cycle logged on a baseline through a vehicle model which can then predict whether or not, under the same real-world conditions as the baseline vehicle, the electric vehicle would have enough range to complete the duty cycle. The charging capability of the vehicle is assessed in a similar way. In order to determine if the vehicle has enough time at night to charge, the simulated vehicle is run through the duty cycle of the baseline vehicle for each day. Analysis determines if there is sufficient time between the last trip of the day, and the first trip of the following day for a vehicle to charge. Based on the analysis of range capability, charge capability, fuelling considerations and other operational costs, a score is assigned to each simulated vehicle. A number of electric vehicles may be simulated, and a 'Best Fit' vehicle is selected from the simulated vehicle that attains the highest score.

2.1.4 Electric Vehicle Monitoring

While this paper focuses on the use of personalized fuel efficiency results in supporting electric vehicle purchases, there is merit in introducing post-purchase support tools. The primary reason this is introduced within this paper is that the positive experience of early electric vehicle owners will have a positive impact on subsequent electric vehicle adoption through word-of-mouth promotion. Given that electric vehicles have real-world efficiency characteristics that differ significantly from conventional vehicles, supporting a first-time electric vehicle owner through the "on-boarding" process is critical. The use of an electric vehicle data-logger that provides real-world efficiency results and explains the factors impacting that real-world fuel efficiency is key.

FleetCarma's C5 data logger was designed to log signals from electric vehicles. Other loggers on the market, while able to provide information on vehicle position and speed could not access signals relating to the electric vehicle powertrain, such as battery state-of-charge, or power used while driving and charging. FleetCarma's C5 data logger records these signals, and transmits them to an online web portal, which then analyses vehicle data providing key metrics for fleet managers, researchers, utilities and other organizations collecting real-world data on electric vehicles.

Electric vehicle monitoring provides several insights for fleet managers. Accessible through a web portal, metrics show the vehicle's utilization and distance travelled, driver behaviour metrics, and charging information.

On each day the vehicle is used by the fleet, the system estimates the available range for the vehicle. This is calculated using charging done at night (bulk charging) or in between trips throughout the day (opportunity charging). Daily range estimates are dynamically changing, based on several factors such as the average daily temperature, auxiliary loads for each trip, and driver behaviour including acceleration and braking behaviours.

Included in the web portal is a fleet-wide report to allow fleet managers to aggregate all their EV utilization metrics and benchmark their fleet's performance against other fleets managing electric vehicles.

The system is also capable of supporting smart-charging projects to reduce grid impact; however, that is outside the scope of this paper. The intent of introducing electric vehicle logging is to support electric vehicle owners after they have purchased to increase positive word-of-mouth promotion of electric vehicles and further acceleration of sales.

2.2 Methodology

2.2.1 Electric Vehicle Modelling and Simulation Process

Electric vehicle modelling and simulation is conducted by installing a small data logger into fleet vehicles. These vehicles are driven for a period of approximately 3 weeks to collect a sufficient amount of data on their duty cycle including any routine variation in the requirements of the fleet application. This information is used to create a costing and emissions baseline that can be referenced when reviewing the predictive performance of new vehicles.

The predictive performance of several electric vehicle models is obtained by using the data gathered from the baseline (existing) vehicle to drive the simulated models of EVs in the same duty cycle. In addition to generating the fuel and electricity consumption of plug-in vehicles the analysis predicts the range and charge capability of electric vehicles completing those duty cycles. As part of this process, the total cost of ownership for the baseline vehicles are compared to the duty-cycle-specific costs of owning and operating EVs doing the same jobs.

2.2.2 Electric Vehicle Monitoring Process

In-service performance monitoring is accomplished with a FleetCarma C5 data logger. The data logger collects information on the vehicle's mileage and utilization, fuel and power consumption, charging information, and driver behaviour. This data is uploaded by fleet managers into an online web portal.

Data from the logger is processed by FleetCarma's back-end system and key performance metrics are provided in the web portal and used to generate a report for each vehicle.

2.3 Accuracy

For a model to be added to the production system it must successfully pass the validation tests. The validation tests include both a trip-based accuracy criteria and a cumulative accuracy criteria. The trip based criteria is that energy consumption is within 10 %, and the aggregate energy consumption must be within 4 %.

3 Results

By generating personal reports and results on a case-by-case basis an increase in the uptake of electric vehicles has been observed. In multiple cases, fleets and individuals who initially expressed an aversion to electric vehicles became strongly interested in purchasing an electric vehicle. It has often shifted the default mindset of "why should I buy an EV?" to "is there any reason I can't?"

The following are four case studies which showcase the effect personalized purchase assistance has on EV adoption. These case studies highlight the key findings from over twenty fleets that have adopted the system. They are a representation for the use of the system for fleets of various sizes, spanning significant geographic and climactic areas.

3.1 A City in Eastern Canada

In 2013, FleetCarma undertook a fleet electric vehicle suitability assessment for a city located in Eastern Canada. The fleet consisted of 40 light-duty vehicles, 20 of which were selected for phase one of the fleet assessment.

The data loggers were preconfigured for the vehicles by FleetCarma, and the fleet team was able to clip them into all 20 vehicles within 1 day. The vehicles were driven for 15–16 days, during which time drive cycle data was captured for each vehicle.

Following this 2 week logging period, the data was uploaded to the FleetCarma servers, where software drove virtual electric vehicle models with the fleet specific data. A summary of the results were then presented.

In one specific case, the vehicle logged was a 2010 Toyota Camry Hybrid, used as a property services vehicle. Over 16 days of logging, the vehicle was on for 10.9 h, and travelled a total distance of 418 km, with the longest distance travelled in a single day of 65 km. The vehicle achieved a fuel economy of 8.0 L/100 km, producing 244 g/km of carbon emissions. The annual total cost of ownership was calculated to be $5,785.[1] The vehicle's utilization is visualized in Fig. 1.

[1] All currency in Canadian Dollars (CAD).

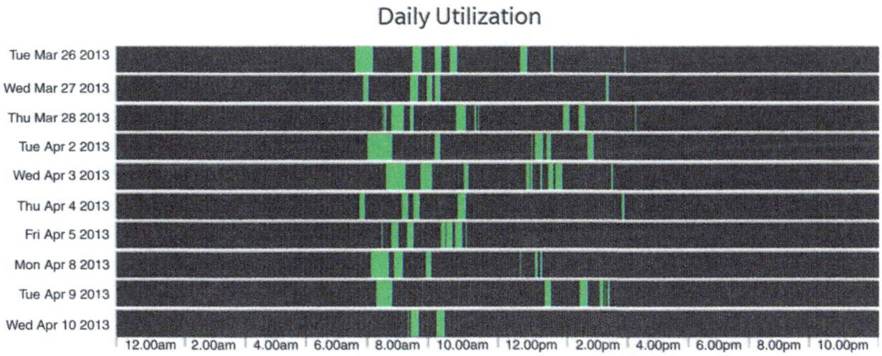

Fig. 1 Benchmarked duty cycle of the 2010 Toyota Camry Hybrid

The figure shows the benchmarked duty cycle's utilization. Each coloured bar represents 1 day with 24 h shown at the bottom. The green portions of the bar highlight times when the vehicle is on, the grey portions show which times each day the vehicle is off.

The benchmarked duty cycle shows routine vehicle usage, with an adequate amount of charging time available between the last trip of each day and the first trip of the following day. This utilization profile, combined with additional factors surrounding fuel consumption, maintenance, and other total cost of ownership variables makes a compelling case for the success of an electric vehicle in this duty cycle.

The vehicles that were being considered to replace the Camry Hybrid were the 2012 Mitsubishi i-MiEV, 2012 Nissan Leaf, 2013 Ford Focus EV, 2012 Chevrolet Volt, and 2012 Toyota Prius Plug-in.

After running the Camry's drive-cycle data through the simulation models, the 2012 Mitsubishi i-MiEV was determined to be the most effective vehicle to replace it, having the highest FleetCarma score. This score considers all of range capability, charge capability, energy usage, emissions, and annual costs.

Based on this specific drive cycle, it was determined that replacing the 2010 Camry Hybrid with a 2012 i-MiEV would result in full range and charge capability, an energy reduction of 85 %, an emissions reduction of 89 %, and an annual Total Cost of Ownership of $4,751; representing an annual saving of greater than $1,000. While each of the four other vehicles considered would drastically reduce energy and emissions as well, only the 2012 Nissan Leaf would also provide an annual cost saving (less than that of the i-MiEV's).

With this information, there were a few questions for consideration: knowing the economic and environmental benefits of changing this vehicle from a hybrid to an all-electric vehicle, could the fleet remarket this Toyota Camry cost-effectively? Could this 2010 Camry be placed on another job, perhaps replacing the 2005 Ford Focus that is doing a building inspections job?

If the fleet could find a suitable replacement application or remarket the vehicle effectively, the model predicted a financial benefit of $6,958 in savings over the

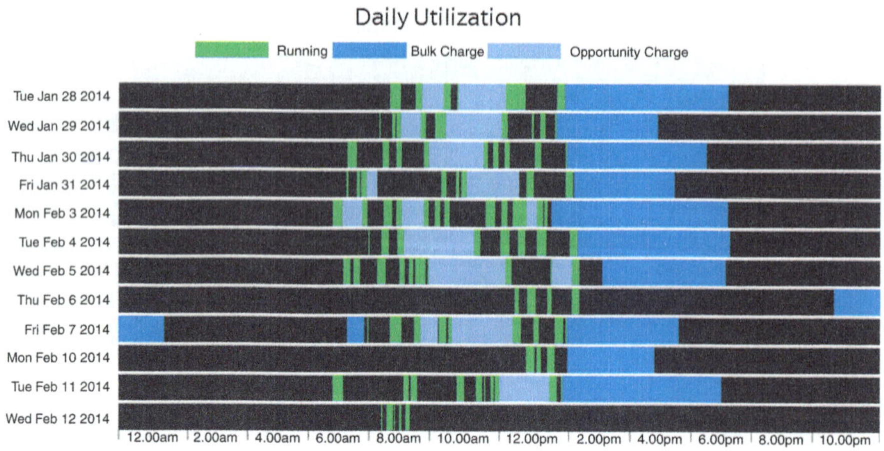

Fig. 2 Duty cycle of the procured 2012 Mitsubishi i-MiEV

7 year ownership life, life-cycle GHG emissions reductions of 12,572 kg CO_2e, and a fuel reduction of 5,664 L of gasoline.

The fleet ultimately decided that the business case for the i-MiEV was too strong not to act. Accordingly, the fleet went ahead and purchased the i-MiEV. The Camry Hybrid was then used in-place of a 2005 Ford Focus, which had also been up for replacement.

Following the purchase, the fleet chose to utilize the FleetCarma electric vehicle monitoring system to optimize the use of the vehicle. In the 4 months the vehicle has been monitored, it has travelled an average distance of 57 km each day, has a 'fuel economy' equivalent to 1.8 L/100 km, and on average ends its day with 41 % of the battery remaining. These results match the predicted results closely. The vehicle's utilization is visualized in Fig. 2.

After the Mitsubishi i-MiEV was purchased and placed in the same application as the Camry Hybrid, the performance of this vehicle was logged to manage and record the success of the electric vehicle in this application.

This figure uses coloured bars represents to represent 1 day with 24 h shown at the bottom. Green portions of each bar indicate the times when the vehicle is on, and the grey portions indicating when the vehicle is off. This graph also includes the time the vehicle spends charging, indicated by light and dark blue. The light blue charging represents opportunity charging, charging that occurs between trips throughout the day, where the dark blue indicates charging that takes place following the last trip of 1 day and the first trip of the following day.

These utilization profiles demonstrate that the vehicle's usage was not impacted by replacement with an electric vehicle.

From a fleet-wide perspective, it was found that out of the 20 vehicles that were data-logged, an electric vehicle would be more cost-effective in 15 of the duty cycles. If the fleet were to adopt these 15 recommended "Best-Fit" vehicles, over the vehicles' life-cycle, the fleet would see benefits of: 21 % in financial savings, a total of \$166,958; an 87 % decrease in fuel consumption, avoiding 160,128 L of fuel; and an 82 % decrease in emissions, reducing their carbon footprint by 465,827 kg of CO_2e.

While the fleet has not immediately purchased all 15 recommended EVs, the first purchase of the i-MiEV may be the most significant. Their experience with the first electric vehicle will lead the discussion when they consider further expansion of their EV portfolio. Had the fleet purchased the wrong electric vehicle for their application or deployed an EV in an ill-suited application, the fleet would be significantly less interested in expanding their EV portfolio. Accordingly, the fleet experience of their first electric vehicle is of paramount importance given its impact on larger-scale integration of electric vehicles into the fleet.

To further build the case for personalization, this fleet EV suitability assessment also offered an interesting scenario. A 2012 Chevrolet Volt was simulated for three different duty cycles which gave a substantially different Total Cost of Ownership (TCO) for each duty cycle. For each of the three cycles, the Volt was predicted to have a TCO of \$43,260, \$50,902, and \$53,221 respectively. What this result demonstrates is the sensitivity of deploying the Volt in different situations. Specifically, although these scenarios had similar annual kilometres driven, the daily distribution of the trips resulted in a \$10,000 difference in the cost of ownership.

3.2 A University in Western Canada

FleetCarma also performed a fleet EV suitability assessment for a university in Western Canada, where 29 light-duty vehicles were logged. The process was similar to the previous case study, where the loggers were all clipped in within 1–3 days, the vehicles drove for 21 days capturing drive cycle data. The logged data was uploaded to the servers where a summary of the results were presented.

In this case study, a 2005 smart fortwo was logged. This vehicle had been utilized infrequently, averaging only 7 km a day, with the longest single daily distance logged being 26 km. The vehicle was consuming 8.9 L/100 km, and producing 275 g/km of carbon emissions.

With this duty cycle, it was evident that all battery electric vehicles were capable of travelling the required distance. The baseline vehicle did not travel far enough each day for the electric range or available charging time to be of concern.

Having such small daily utilization meant that the payback period may not have been as desirable. Battery electric vehicle adoption is most effective when the vehicle's utilization rate lies in the middle of the spectrum. When utilized too much, EVs begin to run into range and charge limitations. When underutilized, the benefit of their fuel efficiency is not fully realised, resulting in a longer payback period.

After running the simulations for eight potential replacement vehicles, it was evident that only two of the vehicles made sense in replacing the 2005 smart fortwo in this drive cycle; the 2013 smart fortwo electric drive and the 2012 Mitsubishi iMiEV.

Implementing the 2013 smart fortwo electric drive would result in an energy reduction of 89 %, an emissions reduction of 99 %, and cost $2,839 annually; a saving of $212 per year. While the financial benefits may not have been large, this university was also able to use the FleetCarma system to present the environmental case for effectively adopting plug-in electric vehicles in the right applications within their fleet.

Knowing the economic and environmental benefits of changing this vehicle to an all-electric vehicle, the university had to consider whether they could remarket the 2005 Smart Car. If they could, it was predicted that they would experience a life-cycle financial benefit of $1,486 in savings, a life-cycle GHG emissions reductions of 3 tons CO_2e, and a fuel reduction of 1,061 L of gasoline.

From a fleet-wide perspective, it was found that the Best Fit EV was a cost-effective replacement in all 29 of the duty cycles logged. If the university were to replace these vehicles with the recommended electric vehicles, they would have the potential to reduce costs by 39 %, a saving of $392,181; reducing fuel consumption by 100 %, avoiding 212,570 L; and reducing their emissions by 99 %, diminishing their carbon footprint by 650,961 kg of CO_2e.

The fleet has since adopted four additional electric vehicles and is using an EV performance-monitoring system to manage their new EV assets to ensure they achieve their desired utilization goals.

3.3 A Town in Eastern Canada

Similar to the case studies previously mentioned, FleetCarma performed a fleet EV suitability assessment for a small town in Eastern Canada. They operate 23 vehicles in their light-duty fleet, 20 of which were selected for the fleet assessment.

In this example, a 2003 Ford Ranger used as a Building Inspection unit was logged for 24 days, travelling a total distance of 312 km, with 51 km being the longest in a single day. It consumed a hearty 15.0 L/100 km, and produced carbon emissions of 464 g/km.

For this drive cycle there were two cost-effective simulated electric vehicles. The best fit vehicle was determined to be the Mitsubishi i-MiEV. The i-MiEV, with full range and charge capability, would provide a 93 % reduction in energy, a 95 % reduction in emissions, and an annual cost of $3,792; a $350 saving over the baseline vehicle each year.

Realizing the economic and environmental benefits of replacing this vehicle with an EV, the fleet had to consider whether they were able to remarket the 2003 Ford Ranger or not. The fleet had an additional concern, could a truck that is being used for building inspections be replaced with a passenger car such as the i-MiEV?

While the goal of a fleet EV suitability assessment is not to suggest switching from a larger vehicle to a smaller vehicle, the potential benefits can sometimes make a case of their own. By replacing the Ranger with a smaller i-MiEV, this fleet would realize a life-cycle financial benefit of $2,467, life-cycle GHG emissions reductions of 14.5 tons CO_2e, and a fuel reduction of 15,295 L of gasoline.

After reviewing the savings, this fleet questioned whether a pickup truck was required in this particular application. Because the TCO of an electric vehicle was defined and compared to their baseline, they were able to consider the true cost of owning a pickup truck as opposed to an EV. They ultimately decided that the benefits of the pickup truck did not outweigh the costs in that particular application, and purchased a 2012 Mitsubishi i-MiEV.

'Right-sizing' presents a very interesting opportunity from a fleet-wide perspective. Within the fleet assessed, six duty cycles in particular offered the potential for substantial financial savings. These duty cycles included two SUVs as well as the Ford Ranger in the Building Inspections depot, and three Dodge Ram pickup trucks in its other departments.

If the fleet were to replace these vehicles with a either a 2012 Chevrolet Volt or 2012 Mitsubishi i-MiEV, the fleet could realize $41,232 in savings (an average saving of $6,872 per vehicle), and would reduce their environmental impact by 180 tons of CO_2. This would reduce emissions by 19 % from all the vehicles assessed, equivalent to taking five vehicles off the road.

3.4 MyCarma for Personal Use

MyCarma, a simplified version of FleetCarma intended for personal use, aims to advise individuals of their fuel savings/costs on a new vehicle based on how they drive. An individual walks into a dealership, receives a complimentary logger, and has it installed in their vehicle's OBDII port. After driving for a week, they return to the dealership to receive their personal fuel economy report.

The personal fuel economy report provides several other metrics for the individual driver to quantify their driving habits. The report provides an eco-driver and eco-route score, the daily distance the vehicle travels, the percentage of time spent idling, and an estimated annual fuel cost. The potential car-buyer is also provided with the projected fuel economy of the vehicles that they are considering for purchase, as well as their projected annual fuel costs and resulting fuel savings.

Creating personal reports for each individual has helped to instil a level of confidence in their purchase decision. In this case study, an individual was looking to replace a 2005 Toyota Sienna. Driving an average of 76 km daily, they were looking to improve on the 11.6 L/100 km they were currently getting, as it was resulting in fuel costs of over $4,000 annually.

Willing to downsize a bit, the individual chose the 2013 Toyota Matrix 4-cylinder, the 2013 Toyota RAV4 AWD Limited, and the 2013 Toyota Prius V as their potential alternatives, with the Matrix serving as their top preference.

After running a personal fuel economy report, the individual found that the potential savings were significant. Replacing their current vehicle with the Prius V would improve their fuel economy to 6.4 L/100 km, effectively cutting their fuel costs to less than $2,300 annually, a saving of $155 a month. These savings would work to reduce their relative car payment. If they went with the Matrix, their savings would be only half that amount (at $82), and even less for the RAV4 ($42 a month).

Shortly after using this information to make a decision, the individual sent this email to their salesperson at the dealership: "Tom, we just filled up for the first time. It cost us $40 and we've driven almost 600 km, I thought something was wrong with the pump. Thank you SO, SO much for showing us that a Prius V was such a good fit for us."

This feedback represents a shift from the traditional response following a vehicle purchase in two ways. First, it reverses the emotional distaste towards the salesman to a positive liking. No longer is the individual wondering if they had been fooled or gotten a bad deal, but are rather thanking the salesperson for advising them on which vehicle best suits them. Secondly, the individual no longer experiences a sense of disappointment after experiencing a fuel economy that is different from the sticker label. The individual was able to rationalize their expectations of fuel efficiency before purchasing the vehicle, and was then delighted when it met that expectation.

Establishing realistic expectations and quantifying EV savings with a high degree of accuracy for individuals will go a long way in increasing their adoption for personal use. The benefits and costs of EV ownership can significantly vary on an individual basis, but when these variances are defined on a personal level, the unknowns are reduced, and the purchase becomes a more viable and realistic option.

4 General Conclusions

For both fleets and individuals, taking a personalized approach to purchasing a new vehicle aids in greater EV adoption. With many individuals, electric vehicles are an unknown technology and are often unproven in their particular application. The ideal method for an individual or fleet to assess whether an electric vehicle would work for them is to purchase one of their own and experience how the vehicle performs not only in the real world, but in the situations and applications that the fleet or individual requires.

While some fleets have purchased electric vehicles as part of a pilot project or trial, these trials present a large upfront cost, and are based on assumptions as to where the vehicle will have the most success. This approach could find an electric vehicle in a duty cycle it cannot adequately perform, and lead to the opinion within the fleet that purchasing an electric vehicle for any application is a bad decision. If the electric vehicle is successful in this position, the fleet may base the purchase of several additional vehicles in different duty cycles on its initial purchase, which could again lead to underutilized vehicles, and may not meet the requirements of the fleet application.

The response to offering a personalized approach amongst the fleets has been positive. As the fleet installs the loggers, it provides more engagement from both fleet management as well as the operators throughout the procurement process. Basing vehicle recommendations on real-world performance demonstrates the practicality of the purchase, and provides more of a bottom-up approach to procuring electric vehicles. The operators on the vehicle are aware of the different stages in the procurement process, and that their own vehicle data is driving the recommendation.

When data is presented as a fleet review, the fleet can realistically assess their utilization patterns, and can visualize how an electric vehicle could complete the same duty-cycle. The review also creates the opportunity for fleets to assess each vehicle's role and conduct right-sizing to realize additional savings.

The fleet review handles common objections to electric vehicle ownership such as range and charge capability while presenting a business case which often strongly favours an electric vehicle as a replacement.

Among individuals, personalization of both their driving habits and their vehicle needs helps to present an unambiguous case for improving their vehicle efficiency. The inclusion of their current fuel costs, contrasted with the fuel costs of several vehicles under consideration not only often justifies the expense of an electric vehicle, but puts a burden on the purchaser to justify any other vehicle choice that would result in a higher overall monthly cost.

Personalizing fuel economy labels and finding the best vehicle for each individual provides a positive sales experience. The experience involves each individual and salesperson collaborating to identify and justify the purchase of a vehicle that will work best for the individual purchaser.

Electric vehicle adoption can be accelerated by removing assumptions in the sales process and providing the purchaser with their own real-world information. This can help to remove common objections to electric vehicle implementation by stakeholders that are concerned with the risks associated with investing in new technology. Addressing each of the risks directly, by providing range and charge capability, helps to reassure the fleet or individual that the electric vehicle can meet their needs without sacrifice or lost utilization. Providing fuel and operating costs in a format that is meaningful to the purchaser enables salespeople to communicate the value of electric vehicles to the purchaser, while also enabling the purchaser to justify that value to internal stakeholders. Most importantly, the value in personalizing vehicle procurement and purchasing is that the vehicle selected is likely to be very successful in the application it is chosen for, leading to a positive user experience and greater overall adoption of electric vehicles.

References

1. Greene D (2010) How consumers value fuel economy: a literature review. National Technical Information Service, EPA-420-R-10-008
2. Greene D (2011) Uncertainly, loss aversion, and markets for energy efficiency. Energy Econ 33 (4):608–616

Part V
Case Studies

Business Models for Electric Vehicles: Lessons from the Japanese EV Ecosystem

Claire Weiller and Andy Neely

Abstract In this paper, we explore the reasons for Japan's early success in the EV industry and the challenges it faces in sustaining its growth in the future. In-depth semi-structured interviews with major players in the Japanese EV ecosystem provide substantial data to draw lessons for EV business model innovation. Current barriers to the EV market are also discussed. We address the impact of the catastrophic tsunami and earthquake that hit Fukushima Prefecture in March 2011 on the emerging EV market. Three main business models are analysed in this paper. First, we present the strategies for the development of the EV charging network in Japan from industry and government perspectives. Second, we discuss innovative business models as drivers of the market with two cases of e-mobility services in Japan. Finally, energy service business models such as vehicle-to-home and storage that allow to capture more value from electric vehicles, are discussed as drivers of entry in the EV market.

1 Introduction

The Japanese EV market is one of the earliest and strongest ones worldwide in terms of sales and industry entry. Since 2005, a combination of factors in Japan has led to the second highest levels of EV sales globally (Fig. 1). Innovative OEMs (Nissan, Mitsubishi, Toyota), a proactive electric utility (TEPCO), and leading

C. Weiller (✉) · A. Neely
Institute for Manufacturing, University of Cambridge, 17 Charles Babbage Road, Cambridge CB3 0FS, UK
e-mail: cw451@cam.ac.uk

A. Neely
e-mail: adn1000@cam.ac.uk

A. Neely
Cambridge Service Alliance, University of Cambridge, 17 Charles Babbage Road, Cambridge CB3 0FS, UK

D. Beeton and G. Meyer (eds.), *Electric Vehicle Business Models*,
Lecture Notes in Mobility, DOI 10.1007/978-3-319-12244-1_12

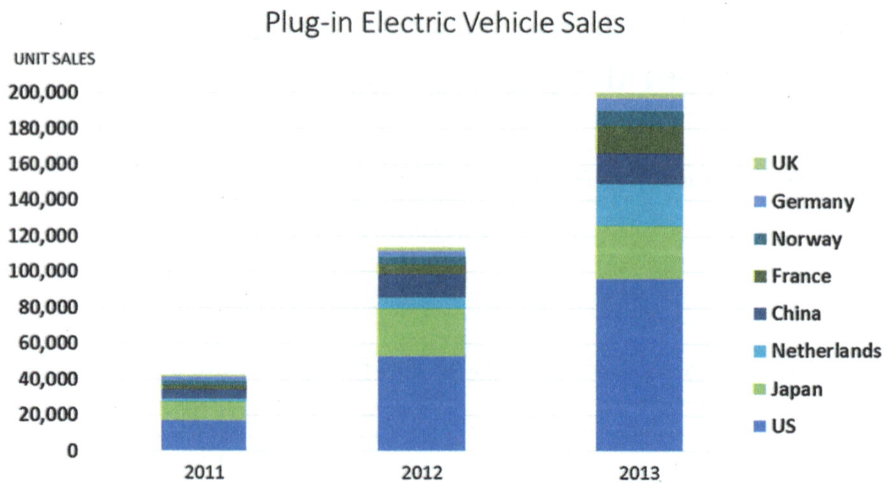

Fig. 1 Plug-in electric vehicle (PHEV & EV) sales in selected countries, 2011–2013. *Sources* Industry websites, http://ev-sales.blogspot.co.uk/

battery and energy companies (NEC, Hitachi, Mitsubishi, Sumitomo) headquartered in Japan have entered the EV market. In addition, a corporate culture that is generally supportive of collaborative R&D across organisations has made Japan a leader in the practice of "open innovation" [1] in many high-technology industries, including automotive and power electronics [2].

In this paper, we explore the reasons for Japan's early success in the EV industry and the challenges it faces in sustaining its growth in the future. In-depth semi-structured interviews with major players in the Japanese EV ecosystem provide substantial data to draw lessons for EV business model innovation. Current barriers to the EV market are also discussed. We address the impact of the catastrophic tsunami and earthquake that hit Fukushima Prefecture in March 2011 on the emerging EV market.

We analyse three main business models in this paper. First, we present the strategies for the development of the EV charging network in Japan from industry and government perspectives. Second, we discuss innovative business models as drivers of the market with two cases of e-mobility services in Japan. Finally, energy service business models such as vehicle-to-home and storage that allow to capture more value from electric vehicles, are discussed as drivers of entry in the EV market.

2 Case Study Data

This case study of the Japanese EV ecosystem is based on data collected from the Japanese EV ecosystem through in-depth interviews with experts from industry, academia, and policy (Table 1). Each interview lasted 1–2 h and followed a

Table 1 Case study interviews

Company or organisation type	Number of respondents	Role or function	Interview reference code
OEMs	4 (incumbents) 1 (start-up)	Firm founder; general managers, strategy planning departments; senior engineers	A1–A4, A5
Electric utilities	2	Senior researchers, R&D department	U1, U2
Battery manufacturers	2	General manager, battery department; senior researcher	B1, B2
Energy equipment and services providers	2	Assistant general manager, battery department; general manager	EQ1, EQ2
Industry experts	3	Research strategy	EXP1–EXP3
Academics	2	Firm founder; research	AC1, AC2
Energy management/software providers	1	Manager, executive, managing consultant, smarter cities department	EM1
Mobility-as-a-service provider	1	Environmental manager	MS1
Management consultancy	1	Consultants	MA1
Automotive research institute	1	Research director, senior chief researcher, FC-EV research division	R1
Engineering design and entrepreneur	1	General manager, design	D1
Governmental ministry	1	Deputy director and director, EV and advanced technology office divisions	GOV1

pre-defined protocol to gain an in-depth understanding of the opportunities and challenges for various e-mobility business models. The interviews were recorded, transcribed and coded in NVivo 10 to analyse content systematically according to the qualitative case study methodology [3].

3 The Development of a Charging Network

In the early days of the EV market in the mid-2000s, the largest public electric utility TEPCO (Tokyo Electric Power Company) started actively promoting the development of the market for EV charging. Through its research, the TEPCO R&D unit for e-mobility recognised the need for the widespread availability of fast-charging to alleviate consumers' range anxiety and to support EV adoption (EXP2).

The TEPCO team, led by expert Dr. Anegawa, initiated the first consortium to develop a common international standard for fast-charging called CHAdeMO. The CHAdeMO charger technology allows a 100-mile EV to fully recharge in less than 30 min at 50 kW (U1). Over 400 companies joined the consortium including RWE, Peugeot Citroen, ABB, and General Electric.[1] Since 2011, the CHAdeMO DC fast-charging connector and standard have been in use in 1,858 stations in Japan and 3,073 worldwide and 57,000 CHAdeMO compatible EVs are on the road as of October 2013.[2]

This success story has, however, been tainted by two major events. First, a competitive threat emerged with Combo, which is a similar standard for DC fast-charging technology developed by German and American automobile manufacturers BMW, Volkswagen and General Motors. While the Combo standard has yet to obtain ISO approval as of 2013, the EV industry largely predicts that it will win this standards battle (AC1, U1) and the European Parliament has already proposed to phase out financial support for CHAdeMO chargers in Europe as of 2018 [4]. Secondly, the activities of the e-mobility team at TEPCO have been severely affected by the nuclear accident in Fukushima. Visionary leader Dr. Anegawa was appointed to the nuclear asset management department to oversee the emergency response to the nuclear plants in Fukushima, and e-mobility research was largely put to a halt after the merger of the unit into the energy storage solutions unit following drastic reductions in TEPCO's R&D budget. The CHAdeMO standard will continue to be manufactured by fast-charging station providers (EXP2), despite the aggressive competition from the US/European Combo standard coupled with the change of focus in the Japanese electric energy industry following the Fukushima catastrophe.

Similarly as in many countries, the government in Japan at federal, prefecture and municipal levels plays an important role in supporting the development of the EV charging infrastructure network. The Japanese Ministry for Energy Trade and Industry (METI) recently announced 1 billion yen ($10 M) funding for slow and fast-chargers (MS1). While the government's role is not to pick technology "winners", it has supported the EV industry through R&D projects in the automotive, ICT and energy sector, particularly as part of smart city projects (G1).

The competitive landscape in the charging network in Japan is characterised by four major industrial consortia that entered the market in 2012: Japan Charge Network, CHAdeMO Charge Network, Nissan Leaf Zero Emissions Service, and the EVSS Network (Table 2). The first one, the Leaf Zero Emissions Support, is an EV after-sale service that integrates all customer needs in IT, maintenance, emergency response, and access to the fast-charging network for a fixed monthly subscription fee. Charge Network Development is a network of fast and slow-chargers in Japan run by the other three main OEMs, Toyota, Honda and Mitsubishi Motors,

[1] CHAdeMO Members list (04/10/2013). Available at: http://www.CHAdeMO.com/pdf/memberlist.pdf.

[2] Website of CHAdeMO association, www.CHAdeMO.com.

Table 2 Industrial consortia for the EV charging network in Japan

	Leaf zero emissions support	Charge network development	Japan charge network	EVSS network
Investors	Nissan	Toyota, Honda, Mitsubishi Motors, Chubu Electric Power	Nissan, Sumitomo, NEC, Showa Shell	Idemitsu Oil, Cosmo Oil, ENEOS, Showa Shell
Background	Comprehensive customer service	Spin-off from CHAdeMO, Membership sharing service model	Nissan dealers, Nexco East Japan, Family Mart	"Next Gen Gas Station" program supported by METI
Number of fast charging stations (Fall 2012)	450	188	22	29
Operations	IT, maintenance, emergency, fast charge	Demonstration service started in Sep. 2012	Started in Oct. 2012 (interoperable with EVSS network)	Started in Oct. 2012 (interoperable with JCN)
Pricing plan	Free fast-charging at Nissan dealers and in some Japan Charge Network stations	Individual members: 1,050 yen/month ($10/month) Corporate members: 3,150 yen/month ($31/month)	Monthly plans from 500 yen ($5) to 4,500 yen ($45), plus charging	Each of the 4 gas retailers have their own pricing plans, including monthly vs. per charge

(*Source* Interview with TEPCO, 02/04/2013)

in association with Chubu Electric Power Company (the third largest electric utility). They are trialling a two-tier membership scheme for corporate vs. individual users on a monthly subscription basis (Table 2). The Japan Charge Network, developed by a collaboration between Nissan, Sumitomo, NEC and Showa Shell, are also testing various pricing schemes with membership or per-usage tariffs (EQ2).

Having a diversity of entrants and of pricing models in the EV charging business is beneficial to stimulate the growth of the market and ultimately increases consumer welfare. The "bottom-up" approach of development of the charging network seen in Japan, where industry partners jointly invest in infrastructure and network solutions and compete for bids for government funding, encourages business model experimentation and innovation with different products and services. Having a common standard for fast-charging, CHAdeMO, enables the rest of the industry to design their own business models in the emerging EV industry. This model of collaborative R&D is found at other levels of the value chain for EVs as well: for vehicle design and manufacturing (e.g. SIM-Drive Corporation), and with joint venture companies such as Automotive Energy Supply Company for batteries and Advanced Energy Company for charging services on Okinawa Island (see Case 1).

In the next two sections, we present innovative business models for EV rental services on Okinawa Island and for energy services.

4 Mobility-as-a-Service Business Models

4.1 Case 1: Okinawa Electric Vehicle Rental Service

In 2010, 200 electric vehicles (EV) were deployed by car rental companies on Okinawa Island (Japan) as part of the "Eco-Resort Island Okinawa Promotion Project". One of the first of its kind, the project included the deployment of a fast-charging network based on Japanese CHAdeMO technology. Through the project, EV rental services are offered by 3 of the islands' many car rental service providers: Nippon Rent-a-car Okinawa (100 cars), Nissan Rent-a-Car Okinawa (60 cars), and ORIX Rent-a-car Okinawa (40 cars). All 200 EVs in the project—which represent 1 % of all 200,000 rental cars on the island—are Nissan Leafs with 160 km range. The energy supply infrastructure for the EVs is provided by Advanced Energy Company (AEC), a company formed in 2010 by a consortium of 26 funding companies with 80 million Yen ($775,000) capital. The main contributing founders include local construction company Kokuba-Gumi Ltd., Nissan Okinawa, and Hitachi Software Engineering Ltd. AEC is responsible for the construction, operations, and services for fast-charging stations (Fig. 2). Alternatively, customers can use the "slow-charging" stations that are available for free on the island and take about 8 h for a complete charge.

The island is reputed for sightseeing and attracts 5.5 million tourists per year, most of which are Japanese, Chinese or other from other Asian countries. About half of the

Fig. 2 AEC fast-charging stations (close-up, *right*; with Nissan Leaf, *left*)

visitors rent a car. Due to its geographical characteristics, the island was thought to be a prime location to introduce EVs in Japan: it spans 130 km from the North Cape Hedo to the Southern tip, and 30 km from East to West. The optimal sites for fast-charging (40–50 kW) stations were determined based on analytical simulations of typical driving routes for sightseeing. Out of 24 proposed sites, property owner and legislative approvals were obtained for 18. A total of 27 fast-charging stations were built on these 18 sites, which can be classified in four categories: gas stations (4 sites), convenience stores, highway service stations, and scenic viewpoint commercial stations.

4.1.1 Challenges of the Okinawa EV Rental Service

High investment costs for an uncertain market In designing the project's business model, due to the high costs of EVs and of charging infrastructure, the investment risk was spread over multiple companies. The following entities participate in the value chain to deliver the EV rental service to customers:

- University of Tokyo researchers, who proposed the project to Okinawan companies and local authorities. Responsibilities include business model design, including revenue structure, long-term market analysis, and research analytics on EV performance and charging sites
- Travel agencies across Japan, who sell package tours that include car rentals from Nippon Rent-a-car and others; bundled booking is the preferred way of renting a car for tourists
- AEC, composed of 3 full-time employees, including general manager Mr. Munehisa Matsumoto (interviewee EQ1), and its 26 founding companies
- Shiraishi Group, the company that owns resorts and the island's branch of Nippon Rent-a-car. Mr. Hiroyuki Nakajima was interviewed as the representative of Nippon Rent-a-car Okinawa (interviewee MS1)
- Alternative "slow" charging service providers, including local companies and tourist sites
- Nissan Rent-a-car and ORIX Rent-a-car Okinawa, other participating car rental companies
- Nissan Motors, manufacturers and providers of the Nissan Leaf cars used in the service
- Local Okinawa authorities and the Japanese government.

Energy provision and regulation Originally, the Okinawa Electric Power Company participated in the strategic discussions concerning energy charging for the EV service. The island is powered by a 26 MW diesel plant and a 10 MW gas power plant. When AEC was created in 2010 and decided to use fast-charging technology rather than standard power infrastructure, the utility retired from the project. Because investments in EV charging stations only benefit EV users, who represent a very small fraction of the Okinawa Electric Power Company service area, regulatory authorities prevented the utility from spreading the costs of the EV infrastructure investments over all of their users, which is the usual way utilities cover their fixed costs. Therefore, the utility did not have a viable revenue model that would allow them to make up for the high capital expenditure for the fast-charging stations. Also due to regulations in Japan governing electricity sales, AEC had to position itself as an infrastructure provider rather than an electricity provider selling kWh. This determined the flat-fee structure of its pricing: AEC charges its customers a fixed price of 2,000 Yen ($20) per rental, which was decided based on the assumption that the average rental lasts 3 days. AEC estimated their business, including investment costs of $8,000 per fast-charging station, would break even based on 10 users per day per charging station.

Customer response The main feedback from customers was that fast-charging stations were too sparse and that the service was not price-competitive with regular ICE rental. 60 % of customers said they feared there were not enough charging stations and that they were anxious about running out of fuel. The density of EV charging stations on the island is 1/10th of that of gas stations, and this is the reference EV drivers use to compare with (MS1).

The car rental companies' objective of reselling the EVs to private users after 3 years failed for lack of demand from the local population. The average income on the island is too low to create an EV market, even for used cars. Customer surveys showed that the willingness-to-pay for the used 3-year old Nissan Leafs is 1.5 MYen ($15,000) rather than the 2MYen ($20,000) the company wanted to resell for. The rental companies have therefore been operating at a loss.

4.1.2 Outcomes of the Okinawa EV Rental Service

The EV rental project was designed as the first stage of a 3-phase "Green New Deal" for Okinawa Island. The long term objective of the project was to develop a smart grid that integrates used EV batteries for electricity storage, an increased amount of wind and solar energy production, and smart home energy management systems. As of 2013, 2 MW of battery storage have been installed by AEC and 1 MW of solar plant capacity has been built at Kanucha Bay Resort as part of the project. The project is therefore well underway towards its initial objective of promoting cleaner, integrated energy systems on "Smart Island Okinawa".

In terms of market uptake, however, the EV rental service did not meet the early expectations of its founders. The "utilization rate", calculated as the proportion of time a car is rented in the year, reached only 10.6 % in 2012 and 20 % overall since the start of the project. For the project to be profitable, an 80 % utilization rate would have been necessary. Three suggestions that came out of the research and stakeholder interviews are presented in the next section. Designed to get the Okinawa EV service back on track towards its financial targets, these recommendations also provide more general considerations for other EV rental services.

4.1.3 Lessons for EV Mobility Services

In the medium term, three main strategic recommendations may help improve the uptake of EVs for rental in Okinawa.

The first solution, which is in the process of being implemented, is to improve customer experience through intelligent route planning or "smart navigation". Through smart tablets and connected devices, intelligent navigation systems could provide highly accurate predictions of energy usage and suggestions for charging stations along the journey based on the customer's driving style and the different types of roads (highways/rural/city). Customers' fear of running out of electricity or "range anxiety" was cited as the biggest barrier to adoption the service, which such planning services can contribute to alleviating.

The second recommendation is to target travel agencies, the weak link in the value chain. Travel agencies, whose services are used by a majority of the tourists in Okinawa, have no incentive to recommend EVs as rental vehicles for their customers. Concretely, during the selection of rental vehicles from the travel agencies, EVs are advertised as a "green" option that curious or environmentally-conscious travellers may use. However, the EV rental started off slightly more expensive than ICE, and even though the cost has since been equalised with ICE, few customers are willing to take the risk. We suggest that rent-a-car companies in Okinawa should identify travel agencies that are willing to include an EV as the *default* car in their holiday package. During the booking service, customers could opt out but otherwise, would be automatically assigned an EV. The default assignment, even if rejected, would at least prompt customers to consider the option in more detail, resulting in a higher probability of acceptance.

The third recommendation is to "open the data". One of the main barriers to acceptance of EVs is the lack of real data on EV usage and experience. The flow of information between the stakeholders in the EV ecosystem must be improved and become more seamless. There is a need for collaboration and information sharing to resolve the practical issues and customer concerns when deciding to rent an EV or not. Combining the data collected from the various stakeholders, from Nissan Motors, the car rental companies, AEC, the tourism industry, and academic researchers, could significantly help refine and improve the service business model. In particular, it would help address customers' questions and concerns at the time of deciding whether to rent and EV or not.

4.2 Case 2: E-Mobility Services in Smart City Projects (Kashiwa and Toyota-City Trials)

Kashiwa and Toyota-city are examples of demonstration phase trials of innovative transport solutions in Japan. Near Kashiwa campus in Chiba prefecture, a joint project between the city authorities, Chiba prefecture, and a development company Mitsui, called upon engineering design manager Ichiro Hatayama to design an optimal transportation system that responds to residents' travel needs while minimising costs and environmental pollution. Through his company Tokyo Design, Ichiro, who has 30 years of experience in automotive design including EVs, tackles the third point of the triple agenda for the smart city: smart energy, an ageing society, and smart mobility.

The resulting vision is to develop a system of autonomous (self-driving) public electric taxis that operate throughout the city and can be called upon at any time to pick residents up at any location. This mobility-on-demand system would allow residents to make all their local trips by car, such as between the home, grocery stores, schools and other local activities, without having to purchase, maintain, and drive their own private cars. For many small and medium city dwellers, as well as an increasing ageing population in Japan, this mobility service addresses their day-to-day driving requirements without the inconvenience of relying on sporadic public transportation. As for the charging ecosystem, the idea in this project is to use wireless charging during the EV operation. The city would not need any traffic lighting as the autonomous driving system in the vehicle would replace traffic controls (D1). Electric vehicles, with their simplicity of use for driving and charging, low operating costs, and environmental advantages, are the ideal vehicle technology for such a service.

Toyota city is another example of mobility service experiments within a collaborative urban planning concept. Toyota city developed around Toyota headquarters and main factory in Aichi Prefecture, Japan. Toyota Motors, in cooperation with the Toyota City municipal government, Hitachi, local public transportation companies, and Chukyo University, developed a small-scale demonstration of a multi-modal optimised urban transport system using EV sharing (see Fig. 3). The service that began in October 2012 provides EVs for its 100 members at four locations in the city for use in conjunction with other public transportation modes. The system offers a route planner for smart phones that takes into account traffic congestion and emissions of different routes, including regional electric power mix for the EV travel. The EV sharing service is meant to fulfil the "last-mile" needs of users who take public transportation into the city and use the car to get from the station to their final destination. Toyota will use the trial to collect data on EV battery usage in an energy data management system. In the early days, the system is provided for free at four locations, two at the partner University and two at local

Fig. 3 The Ha:mo EV-sharing station in Toyota City (*left*) and the Ha:mo Navi smartphone application (*right*)

railway stations. Plans are to increase the number of members to 1,000 and increase the number of stations to 20. As part of the Next-Generation Energy and Social Demonstration projects sponsored by the Ministry of Economy, Trade and Industry, the tests are conducted by Toyota City's Low Carbon Verification project. The aim of this collaborative trial is to lower overall energy use from transportation in the area and improve the efficient of transport for users' needs.

The realisation of such innovative projects requires large scale investment and behavioural changes, which only collaborative efforts between the various stakeholders can enable. In the Japanese case as in other urban ecosystems in the world, the important players include businesses such as information and communication technology (ICT) solutions companies, academic and research institutes, entrepreneurial firms and investors, and government and utilities [5]. Partnerships ensure that all of the elements are provided: capital investments, human and technological resources, trial participants (e.g. University students or company employees), and longitudinal (time) resources to monitor the evolution of the project.

For entrants in the EV ecosystem, and particularly in the EV charging business, a major question of designing an e-mobility business model is how broad to extend the scope of the offering. Companies can choose between developing technical standards for the hardware and connectors, as CHAdeMO did (Sect. 1), or designing infrastructure and network services, or embedding charging services within mobility service platforms, i.e. focusing on the whole driving experience, as in the case of Okinawa Island (Case 1). Finally, even broader platforms integrate electric mobility and their charging systems as just one piece of the puzzle, such as smart home energy management systems and smart cities (Case 2).

5 Energy Service Business Models

Throughout these interviews in Japan, it became clear that many companies entered the EV industry with a longer term objective to develop "smarter", cleaner energy systems and technologies that can be re-used in the broader market for energy management services, such as battery and storage systems. As described in the case studies of Okinawa and Toyota City, the transition to mobility services with EVs is often part of a wider transition to smarter energy management systems at the level of the home, the grid or the city.

For battery manufacturers and energy equipment providers, such as NEC, Hitachi, and Sumitomo in Japan, the development of lithium-ion batteries for EVs offers multiple sources of value: first, the opportunity to enter a new growth market (EVs), and second, the opportunity to re-use the knowledge and technology developed for the EV market in other markets and in broader applications. For some firms, the EV market clearly represented the starting point to new business in other energy markets: the "energy storage business, such as home energy storage or residential storage and community storage" (B1). This can be called the technology "spill-over" effects. Sumitomo, a trading company involved for over 15 years in trading raw materials in the value chain of EV batteries including lithium, nickel, and cobalt, is conducting research in EVs and energy storage markets in view to broaden the battery market and open new markets for these raw materials.

One concrete example is the formation of a joint venture company between Sumitomo (49 % share) and Nissan (51 %), 4R Energy, for the development of batteries for home energy management services. Given the capabilities and stock accumulated by Nissan of batteries for the Leaf, 4R Energy was set up as a joint venture company in 2010 to develop a new market for the used batteries (EQ2). The batteries come from the same manufacturing facilities as the 24 kWh batteries made for the Leaf by Advanced Energy Service Company (AESC), another joint venture between Nissan and NEC (EQ2). After their life in the vehicle, the batteries are recycled into two 12 kWh batteries and sold on the retail market for use as electricity storage systems in households (A1, EQ2). In Japan, the market for domestic battery solutions has gained traction after the shutdown of the Fukushima nuclear plants which cut off power supply for hours and raised concern for energy independence from the grid among the population. 100 units of the 4R battery have been sold in the first year of operation (EQ2). The price of the service package including the battery, maintenance, installation and all customer service is 3 million yen. While this is still very high, the price is expected to decrease as the product diffuses in the market (EQ2).

Moving into the domestic energy storage market through 4R Energy and into the business for EV charging through its partnership in Japan Charge Network, Sumitomo entered the EV business to open new market opportunities and be at the forefront of any growth market in the battery value chain from raw materials to end-user services.

4R Energy is not the only company of its kind. A competitor, ORIX Corporation, recently established a company that provide battery services for residential houses. For 3,000 Yen per month, the company offers 6 kWh batteries and the ICT management system to optimise its use (EXP3). NEC also sells 6 kWh residential use batteries and other 1–2 kWh smaller ones are available on the market (EQ2). While the initial goal is the introduction of residential batteries, the ultimate vision is to control the power from 1,000 to 2,000 batteries at once through cloud services (EXP3). ORIX is a financial trading company and it would like to see a market take shape to trade electricity between power companies with the home batteries' energy (EXP3).

This vision of aggregating EV battery resources to exchange power with utilities and grid-level players is often discussed in the literature on EVs as the "vehicle-to-grid" (V2G) concept [6, 7]. Through the interviews in this research, we found that V2G will require a more complex ICT infrastructure to be deployed between home energy management systems and grid controllers, which is currently not justified by the low penetration of EV batteries. We expect V2G to be on the commercial horizon in the long term (10–15 years) [8].

As the lithium-ion battery technology improves, larger grid-scale energy storage systems are expected to be released on the market in Japan as well. Such grid mega-storage systems of 1.5 MWh are currently in demonstration and testing phases with utilities (EQ2).

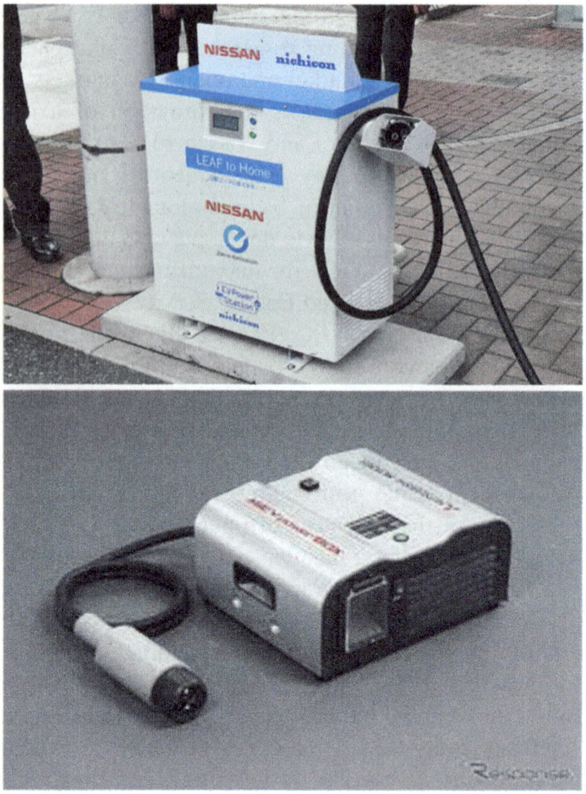

Fig. 4 Nissan Leaf-2-Home (*top*) and MiEV Power Box (*down*): two vehicle-to-home devices in the Japanese market

Automotive manufacturers have also started to take an interest in the energy services associated with the electrification of vehicles and have been seen to move into the business of energy solutions in Japan. Mitsubishi's MiEV Power Box and Nissan's Leaf-2-Home system are examples of devices that enable the transmission of electricity from EVs to the home (Fig. 4). Nissan Leaf buyers can currently purchase the Leaf-2-Home for about €4,000 (U21) to use the car's 24 kWh battery as a source of energy for their homes. The MiEV Power Box, which is also on sale in Japan, offers 16 kWh energy capacity for home use (A24). The Leaf system could potentially provide up to 2 days of electricity for an average Japanese household (R1, A1). These systems of home energy supply from EVs were tested and deployed following power disruptions after the earthquake and tsunami in 2011. Gasoline supply was interrupted for a month in the region, whereas electricity infrastructure was quicker to recover (A4). Mitsubishi delivered 90 iMiEVs to the Prefecture at that time for use as home batteries. Customer feedback was positive and proved the value of such vehicle-to-home systems.

In the case of Okinawa Island's EV rental service, the EV business was only the first phase of a longer term strategic transition to "green" the island's energy sources with smart grid technologies, renewable energy generation, and efficient buildings (EQ1). EV batteries were intended to be recycled after their life in the EVs and reused in buildings and homes for energy management and storage services. These batteries were seen as an important component of the system to balance the grid at times of excess or undersupply of energy. EVs themselves, of course, also provide a solution to improve the environmental sustainability of the island.

In summary, three pathways for value creation and capture in energy services business models with EVs have been found in Japan: the technology spill-over effect of EV battery technology into other markets and applications, the secondary life value of EV batteries, and the direct use of EV batteries in the vehicle for energy management in the home.

At the moment, OEMs are bearing the largest share of the investments and costs in EV technology (EQ2, A1). The results in this paper suggest that risk-sharing participation from companies in other sectors such as electricity supply and equipment can enhance the viability of the market and create business opportunities in the emerging EV ecosystem.

6 Conclusions

The case of the Japanese EV ecosystem contributes strategic perspectives from multiple points in the value chain for EVs, from battery and car manufacturers to mobility-as-a-service providers (Fig. 5). Examining the business models that led Japanese companies to have a global presence in the EV market has provided three significant insights as to how to create and capture value in early EV commercialisation.

Firstly, the provision of a seamless charging network is an essential part of the value proposition for EV customers. In Japan, a combination of government investment and corporate investments have led to the formation of consortia of companies developing the charging network with both fast- and regular chargers. The first standard for fast-charging connectors was developed by the CHAdeMO association initiated by TEPCO to address the problem of recharging wait times. Defining a financially profitable business model in the early stages of the EV market, where sales are still low, is still a challenge. The consortia in Japan operate on a shared risk model (joint investments) and on an experimental basis where different membership schemes are available and charging stations are trialled in

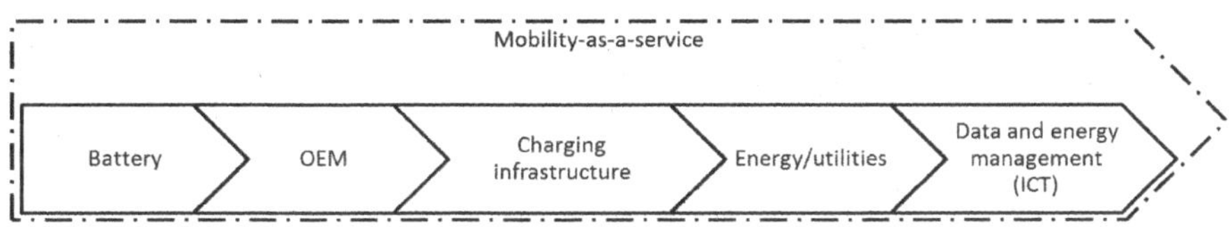

Fig. 5 A basic representation of the EV value chain

preferred locations such as tourist sights, participating shops and gas stations. The deployment of a charging network as a competitive, innovation-driven process is beneficial to the growth of the industry.

Secondly, a competitive advantage for firms in the EV market can be obtained by investing in technologies and competencies. However, a strategic recommendation from this paper is to have a "smart" business model that uses digital technology to address end-user needs. Innovating a business model can be just as powerful as acquiring technological advantages to compete in the EV ecosystem [9]. The cases of the EV rental service and of e-mobility in smart cities in Japan highlight the importance of overcoming the barriers to EV adoption by formulating business models around the *end-user driving experience*. Firms in the EV ecosystem must take a customer-centric view to design the right end-user experience.

Finally, the business model for EVs must give customers more value from their car. In Japan, the potential for mobility services is limited, but energy services business models such as integration in smart home systems, smart cities, and the reuse of batteries for grid management applications, are new sources of value that the industry stakeholders are already starting to tap into.

The Fukushima nuclear reactor crisis following the earthquake and tsunami that hit Tohoku area in March 2011 caused a significant shift in customer perceptions of the electricity industry, making them less accepting of any related innovations such as electric vehicles. It also readjusted the priorities of energy infrastructure and supply companies towards stabilizing mechanisms such as storage batteries, renewable energy and smart grid integration, and away from shorter-term/secondary priorities such as electric mobility. However, the EV is widely seen in Japan as one element of a greater structural transition to a sustainable and efficient "new energy society" (EQ1). One of the drivers of the EV market in Japan are the long-term strategic and financial value creation opportunities discussed in this paper. Such opportunities may inspire new business models and entry in other markets globally to stimulate competition and create a viable EV ecosystem [10, 11].

References

1. Mortara L, Minshall T (2011) How do large multinational companies implement open innovation? Technovation 31(10–11):586–597
2. Christensen J, Olesen M, Kjar J (2005) The industrial dynamics of Open Innovation—Evidence from the transformation of consumer electronics. Res Policy 34(10):1533–1549
3. Eisenhardt KM, Graebner ME (2007) Theory building from cases: opportunities and challenges. Acad Manag J 50(1):25–32
4. European Parliament Committee on Transport and Tourism (2013) Draft report on the proposal for a directive of the European Parliament and of the Council on the deployment of alternative fuels infrastructure
5. Visnjic Kastalli I, Neely A (2013) Collaborate to innovate: how business ecosystems unleash business value. Cambridge Service Alliance Executive Briefing, Cambridge
6. Kempton W, Tomi J (2005) Vehicle-to-grid power fundamentals: calculating capacity and net revenue. J Power Sources 144(1):268–279

7. Quinn C, Zimmerle D, Bradley TH (2010) The effect of communication architecture on the availability, reliability, and economics of plug-in hybrid electric vehicle-to-grid ancillary services. J Power Sources 195(5):1500–1509
8. Weiller C, Neely A (2014) Using electric vehicles for energy services: Industry perspectives. Energy (in press). doi:10.1016/j.energy.2014.06.066
9. Chesbrough HW (2010) Business model innovation: opportunities and barriers. Long Range Plann 43(2–3):354–363
10. Adner R (2012) The wide lens: a new strategy for innovation. Portfolio/Penguin, London, p 288
11. Adner R, Kapoor R (2010) Value creation in innovation ecosystems: how the structure of technological interdependence affects firm performance in new technology generations. Strateg Manag J 31(3):306–333

Orchestrating Ecosystem Co-opetition: Case Studies on the Business Models of the EV Demonstration Programme in China

Tianjiao Shang, Ying Chen and Yongjiang Shi

Abstract This paper explores the development of the Chinese EV demonstration programmes through the conceptual lens of the business ecosystem framework. At present, government is acting as an ecosystem orchestrator promoting different types of business models such that the Battery Swapping Model in Hangzhou and Battery Charging Model in Shenzhen are co-existing while competing fiercely; such actions of orchestrating ecosystem co-opetition triggers practical implications for policymakers and industrial players who are tackling the challenges of stimulating a supportive business environment to promote the advancement of the EV industry in their respective countries. Theoretically, this paper clarifies the definition of business ecosystems, both statically and dynamically, while examining the relationship between business ecosystems and business models.

Keywords Business ecosystem emergence · Electric vehicle industry · Ecosystem co-opetition · EV business models

1 Introduction

Over the past few decades, there has been renewed interest concerning the development of electric vehicles (EV). The emergence of this industry is related to the recent sharp increase in oil prices along with the environmental concerns over

T. Shang · Y. Shi (✉)
Department of Engineering, Institute for Manufacturing, 17 Charles Babbage Road, Cambridge CB3 0FS, UK
e-mail: ys@eng.cam.ac.uk

T. Shang
e-mail: tjs69@cam.ac.uk

Y. Chen
School of Management, National Institute for Innovation Management (NIIM), Zhejiang University, Hangzhou 310058, People's Republic of China
e-mail: chen1212ying@zju.edu.cn

D. Beeton and G. Meyer (eds.), *Electric Vehicle Business Models*,
Lecture Notes in Mobility, DOI 10.1007/978-3-319-12244-1_13

climate change. As a matter of fact, policy makers have come to comprehend the intricate impediments in reducing greenhouse gas emissions (GHG) from the transport sector, especially given the strong linkage between transport consumption and economic growth. As a consequence, the impetus in advancing this emerging industry is strong, with governments across the globe deploying demonstration programmes so as to promote EV development.

The purpose of this paper is to explore the development of the emerging Chinese EV industry and to update the progress of the EV demonstration programmes orchestrated by the policymakers in China. Moreover, the examination of this emerging EV industry is carried out through the conceptual lens of a Business Ecosystem (BE) framework. The employment of the business ecosystem concept is required as the emergence of EVs involves heavy infrastructural support (i.e. the establishment of charging facilities) and the collaboration between diverse sectors of players across industries, especially at the embryonic stage of its advancement. Therefore, business ecosystem is particularly useful for the analysis of this industry compared with alternative concepts such as Supply Chains due the emerging nature of its development. At the same time, it is the aim of this paper to conceptually clarify the definition of business ecosystems, both statically and dynamically, while distinguishing the relationship between business ecosystem and business models.

Employing case studies from the EV demonstration programmes in Hangzhou and Shenzhen, this paper implements cross-sectional ecosystem mapping while identifying the competing and co-existing business models of these EV business ecosystems; aiming to shed more light for the policymakers in tackling the challenges of stimulating a supportive business environment to promote the advancement of the EV industry.

2 The Business Ecosystem Framework

2.1 Business Ecosystem Review

Moore originally proposed the concept of business ecosystem in 1993. He defined the term as "An economic community supported by a foundation of interacting organizations and individuals—the organisms of the business world. This economic community produces goods and services of value to customers, who are themselves members of the ecosystem. The member organizations also include suppliers, lead producers, competitors and other stakeholders. Over time, they co-evolve their capabilities and roles, and tend to align themselves with the directions set by one or more central companies". In Moore's view, the term business ecosystem not only encompasses the core supply chain and extended enterprises but also envelopes other stakeholders such as industrial associations and the government [1, 2]. More importantly, it is stressed by the pioneer of business ecosystem theory that the co-evolution and interaction among these different level of organisations including firms within the supply network, extended enterprises and policy makers are critical for their co-development.

Since Moore's proposal of the business ecosystem concept, many scholars have endeavored to conduct research within this field. Surveying business ecosystem studies since Moore's work, our working definition of business ecosystem is said to be "a community consisting of different levels of interdependent organisations, who are loosely interconnected and generate co-evolution between partners and their business environment" [3–9]. The loosely interconnected feature is emphasised as a number of papers have stressed the importance of many "loosely interconnected" members who rely on one another to survive and acquire mutual advantages within a business ecosystem [4]. Based on Moore's ecosystem, Iansiti and Levien identified four categories of players, which are the keystone player, the niche player, the dominator and the hub landlord who are participating within the ecosystem, such that the functions and strategies of these players are pinpointed. Accordingly, Iansiti and Levien contend that the task for the keystone player is to provide a platform allowing other participants to work and collaborate with one another [5].

Concerning the evolutionary development of business ecosystems, the life cycle of a business ecosystem includes the phases of birth, expansion, authorities and renewal [1]. Adapting from Moore, Rong enriched the business ecosystem life cycle concept with five phases of emerging, diversifying, converging, consolidating and renewing as the evolutionary pathway of the ecosystem using cases studies from the semi-conductor industry [9]. Dynamically, business ecosystems play critical roles in the nurturing of emerging industries; the four building blocks allowing the ecosystem to operate dynamically include its Resources Pool (or Social Network), Interaction Mechanisms, Value Network and the Business Context [10] (Fig. 1).

2.2 EV Business Ecosystem Structure

The business ecosystem encompasses three main sub-systems, which are BE Supply, BE Demand and BE Intermediaries. Through analysing data from the EV industry [10], Fig. 2 depicts its BE structure such that the supply side is concerned with the production of the EVs encompassing core components manufacturing firms such as battery manufacturers, traction motor suppliers, the final assemblers as well as the car body component makers. Meanwhile, the EV users occupy the demand side. BE Intermediaries consists primarily of dealers and retailers such that it provides the opportunity in which supply and demand are integrated. In addition, BE Intermediaries encompasses supporting participants of the ecosystem including research centres from universities, electricity providers such as the state grid, industrial associations and the government.

The business ecosystem concept is particularly useful for the industrial players and governments concerning the development of the EV sector, as the emergence of EVs requires heavy infrastructural support and the collaboration between many

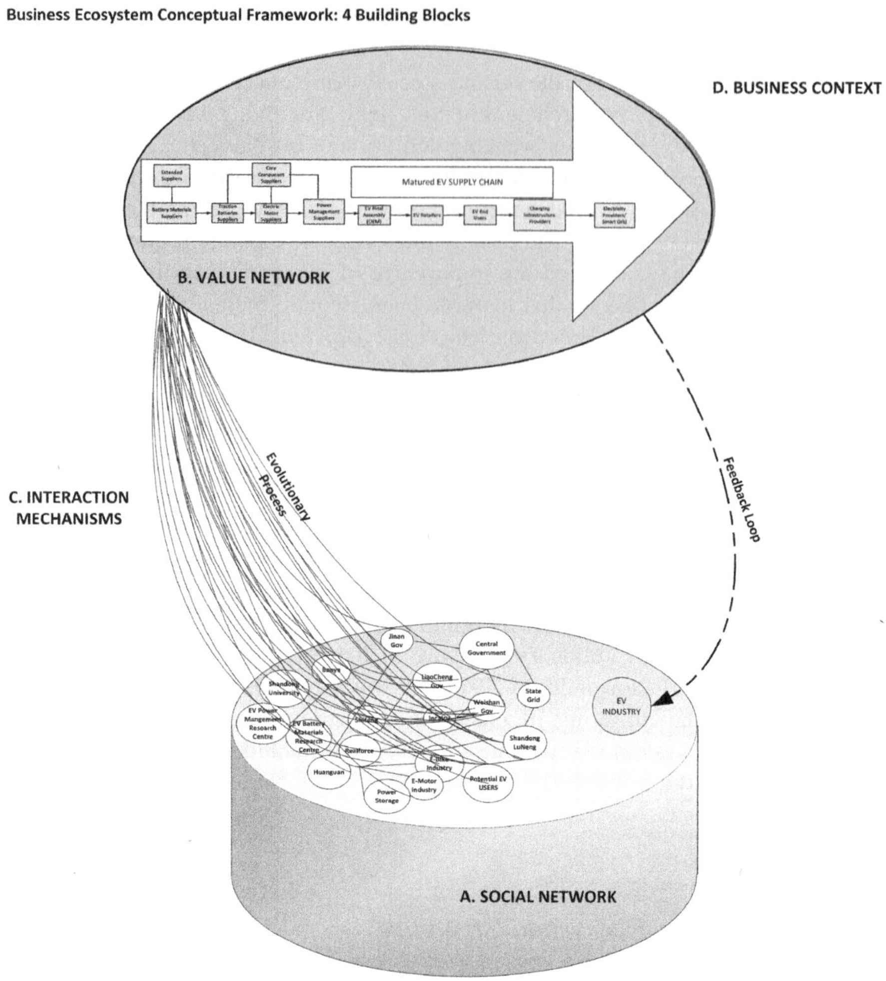

Fig. 1 Emerging business ecosystem dynamics [10]

players across a spectrum of industries (e.g. the establishments of charging points). Furthermore, because of the uncertainty at the emerging stages of the industrial development, policymakers are important ecosystem players providing platforms for the nascent sector to progress. Analysing the case studies from the cities of Hangzhou and Shenzhen, this paper conducts EV ecosystem cross sectional mapping, through which the structure and the mechanisms of ecosystem orchestrators coordinating the EV business ecosystem could be comprehended. More importantly, the value capturing opportunities from competing business models incentivized by the demonstration programmes could also be identified.

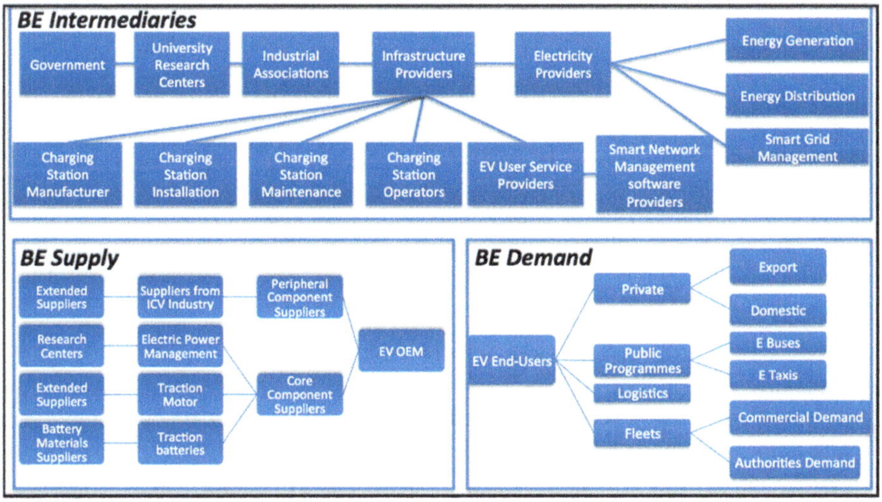

Fig. 2 EV ecosystem cross-sectional structure [10]

3 The Chinese Electric Vehicle Demonstration Programme

3.1 Demonstration Programme Overview

China overtook the U.S and became the largest automotive market in the world following the introduction of sales incentive by the central government in 2009. This rapidly growing market has attracted all the dominant automakers to gain presence within the country and it is projected that the demand for vehicles is to expand further in the long term, as car ownership is still low compared with the developed countries. Accordingly, the Chinese government devotes a great deal of attention to the automobile industry and has set the EV sector as one of its main strategic industries in achieving high economic growth.

China initiated research and development concerning the EV industry at the beginning of this century. Following the implementation of the "Key EV Project" and the "Key project of Energy-saving and New Energy Vehicles" from the National 863 Program, the Ministry of Science and Technology invested approximately 2 billion (RMB) in the course of the 10th five-year plan and the 11th five-year plan. During this period, approximately 500 associated projects were executed including the provision of funds for R&D, the establishment of various industrial standards, the subsidy for the end users and manufacturers of EVs, the promotion to electricity providers of the investment in relevant infrastructures as well as the development of related legislations [11].

In 2009, the Chinese government carried out the "Thousands of Vehicles, Tens of Cities" program. This is an EV demonstration project such that subsidies are given to the 25 pilot cities and EVs will be used in the public transportation systems involving

Table 1 Exploratory case studies overview

City	OEM	Battery supplier	Motor supplier	Intermediaries	Government
Hangzhou	WanXiang Group (*Provider of E-Bus and private EVs*)			Zotye Auto Retailers (*Provider of E-Taxi and private EVs*), State Grid	Hangzhou EV demonstration implementation office
Shenzhen	BYD Auto (*Provider of E-Taxi, E-Bus and private EVs*)			China Southern Power Grid	Shenzhen EV demonstration implementation office

buses, taxies, government vehicles, cleaning vehicles and postal vehicles. Among the 25 demonstration cities, 6 cities were chosen as the pilot cities for both public and private demonstrations of EVs. The prediction is that the Chinese EV market will reach 1 trillion (RMB) by 2020. So far, 217 innovative EV models have been launched by domestic firms and by 2015, the industry aims to achieve a production of more than 1 million. Both cities, Shenzhen and Hangzhou were chosen to conduct demonstrations programmes for public transportation as well as individual EV purchases.

Exploratory case studies were carried out through the implementation of semi-structured interviews as well as on-site plant visits with EV OEMs, policymakers and associated ecosystem players for the purpose of data collection. In particular, the government offices responsible for the implementation of the EV demonstration programme have been visited; interviews were conducted concerning the pilot EV development for both Shenzhen and Hangzhou. Table 1 provides an overview of the data collected across the EV business ecosystems players in Shenzhen and Hangzhou.

3.2 Demonstration Programme: Hangzhou's Battery Swapping Model

The pilot EV demonstration project initiated in 2009 in the city of Hangzhou, as the demonstration city for both public and private transportation, the city is targeting to promote a rental business model for EV users so as to mitigate one of the main barriers which is the high economic upfront cost associated with batteries ownership when purchasing new energy vehicles. WanXiang is one of the main OEMs participating in the EV demonstration programme in the city of Hangzhou. There are two types of EVs supplied by WanXiang: electric buses and private EVs.[1] The electric buses were served as public transportations inside the exhibition areas during the

[1] WanXiang's passenger EVs are not used in the Hangzhou taxi fleet as opposed to the case of BYD. The EVs produced by WanXiang are part of the demonstration programme for private purchases, while the participating OEM supplying electric taxis for the Hangzhou demonstration programme is Zotye Auto.

Shanghai Expo. The private EV model HAIMA has a range per charge of 150 km with a charging time of 3 h and a maximum speed of 110 km/h. The business model of WanXiang focuses on battery swapping and offers a battery rental model. The HAIMA EV manufactured by WanXiang Group can be rented from retailers at a monthly cost (without the battery) while the battery can be rented from the State Grid on a monthly basis (costing around 200 GBP per month). However, during the first 3 years or 60,000 km following the EV purchase, customers enjoy free battery usage and swapping services through a government subsidy. As a result, the entry cost and the usage cost of the EVs have been significantly reduced. In addition, WanXiang is cooperating with the State Grid on developing a standardised EV battery pack that enables a quick system for battery exchange in the stations operated by the State Grid Corporation of China in Hangzhou.

Concerning this key ecosystem player from the BE Intermediaries sub-system, State Grid Corporation of China (SGCC) is a government owned enterprise that aims to provide clean and sustainable electric power to the public in China. The company ranked 8th in the Fortune Global 500 ranking in 2010. The major business of SGCC is the building and operation of electricity grids in 26 provinces serving over 1 billion people in China. SGCC is closely related to the EV industry since it controls the major electricity supply facilities in China. By realizing the importance of the EV industry to the country's economy and also the potential opportunities, SGCC has been investing on the construction of EV charging facilities. Collaborating with EV ecosystem players in Hangzhou, SGCC has established three companies for the implementation of the EV demonstration programme: (1) for the operation of Hangzhou's Electric Taxis (2) for the operation of battery swapping (3) for the battery swapping infrastructure. SGCC is working on a project regarding the construction of centralized battery charging facilities that use cheaper off-peak electricity. At the same time, local battery changing stations where the fully charged batteries can be sent for the use of end-users are being constructed; such that charging points and mobile battery changing trucks would be available around the city of Hangzhou in the future for EV users for emergency needs.

3.3 Demonstration Programme: Shenzhen's Battery Charging Model

Shenzhen supports the battery-charging model for EVs instead of battery swapping for both public and private transportation. The office estimated that 78 public charging stations for private EVs such as BYD's F3DM, and 22,000 charging poles would be constructed before 2013. Quick charging will be available in charging stations while charging poles offer normal charging. The Shenzhen government will also provide subsidies for the private purchases of pure EVs so as to lower the entry barrier. BYD, which is the key EV OEM supplying pilot vehicles for the demonstration project in Shenzhen and has released three EV models, the K9, the F3DM and the E6 since 2005. The K9 is a 12 m pure electric bus with a range of 250 km per charge. The F3DM is a plug-in hybrid EV. The E6 is a five-door hatchback EV used as taxis in Shenzhen's demonstration project operated by a joint-venture company between BYD and China Southern Grid. In addition to the taxi fleets, BYD offers the E6 and F3DM to private customers along with two charging posts for free at the location of consumers' preference. This establishment of charging infrastructure is a result of the collaboration between BYD and the electricity provider, China Southern Grid, in Shenzhen. The business model of BYD does not directly reduces the battery, vehicle and electricity costs for potential EV users, but subsidies have been granted from both the central and local governments in seeking to address such price concerns.

China Southern Grid (CSG) was founded in 2002, providing electricity for five provinces in Southern China. CSG has a power network of over 1 million km^2 serving a population of 230 millions. EV charging technology is one of the main research areas for the enterprise. CSG has successfully constructed the first batch of EV charging facilities in 2002, which are located in Shenzhen city. CSG is cooperating closely with BYD Auto on the EV charging technology project. By creating a joint-venture taxi operation company with BYD, CSG has built two fast charging stations and over 100 charging poles in Shenzhen city for the battery charging of pure EV taxis. Different from the SGCC strategy, CSG provides battery charging services instead of battery changing and thus without the need of promoting a standardized battery package.

4 Competing and Co-existing Business Models

4.1 Business Model Review

The concept of *Business model* gained wide attention since the 1990s following the flourishing development of the Internet; both in academic circles and management practice. Business model is a holistic perspective on how firms conduct business or "create, deliver, capture value" [12, 13]. At the operational level, elements such as customer relationship/information [14, 15], process and activities [16]; information flows [12] have been identified as components of a business model. At the strategic level, business model itself is not strategy [17], but contains strategic elements, such as customer value proposition [18], position in the value chain [19] and product scope [15]. Firms could adopt the same strategy by using different business models and different operational practices accordingly. Furthermore, Hamel [14] connects the strategy elements with the business model by listing four components: *core strategy, strategic resources, value network, and customer interface*. Moreover, Shafer [20] find that there are 42 components used in existing various definitions and classify these into four final clusters, i.e. *strategic choice, value network, create value, and capture value*. Osterwalder identifies nine business model components: *value proposition, customer segments, partners network, delivery channel, revenue stream, relationship, value configuration, capability, cost structure.* In fact, Osterwalder [21] is among the earliest scholars who realize that the match or fit between different blocks' possible choices and the need to align with each other should play a very important role. To implement a business model, the focal firm should consider strategic elements such as who we are aiming at, where is our position in the value chain, how to combine all possible products to meet the needs, as "A business model is a reflection of the firm's realized strategy" [22].

Concerning business model emergence, Björkdahl [23] finds that new technology calls for novel business models in order to promote its adoption, which echoes with the research of Chesbrough and Rosenbloom [19]. Abernathy [24] identified the latent logic behind the transformation of the above regime by using the concept of dominant design. Before the emergence of the dominant design, firms focus on product innovation because this ferment period abounded with the uncertainty about customers' needs, the technology specification. In this stage, the focus of competition is "the nature of the game" [25], e.g. how to promote certain technology as the dominant design afterwards. By using the concept of value net that refers to a firm's linkages with its suppliers and producers of complementary products, Raji et al. [26] have found that the size of the value net will shorten the time to the emergence of dominant design. More recently, Pek-Hooi [27] using the strategic alliances samples in USA showed that central firms with high ego network density, coupled with a strategic intent to acquire and share knowledge broadly within the technological community, will achieve better innovation performance before the dominant design emerges. In summary, at the early stage of an industry, i.e. before the emergence of the dominant design, the lack of knowledge on technology, customers' preferences and the ways in which to incorporate the innovative technology into a profitable business provides opportunities for firms with a novel business model.

4.2 Competing and Co-existing EV Business Models in China

Following the implementation of "Thousands of Vehicles, Tens of Cities" demonstration project in China, it is evident that competing business models have emerged. Snapshots of the EV ecosystems and the mechanisms of the ecosystem orchestrators coordinating the value capturing opportunities from competing business models incentivized by the demonstration programmes have been captured (Figs. 3 and 4). Figure 3 demonstrates the interactions between the EV ecosystem players implementing the battery swapping business model in Hangzhou. This business model effectively incentivises consumers to adapt behaviour, while reducing the risks of battery ownership. The orchestrator has encouraged State Grid to formulate the mechanisms clearly through working in collaboration with WanXiang and Zotye Auto. The business model has diversified from a traditional model through a change of value proposition for its potential users. However, due to the sunk costs in the co-development with other ecosystem players of the battery swapping standardisation and infrastructure, this business model has the disadvantage of inflexibility.

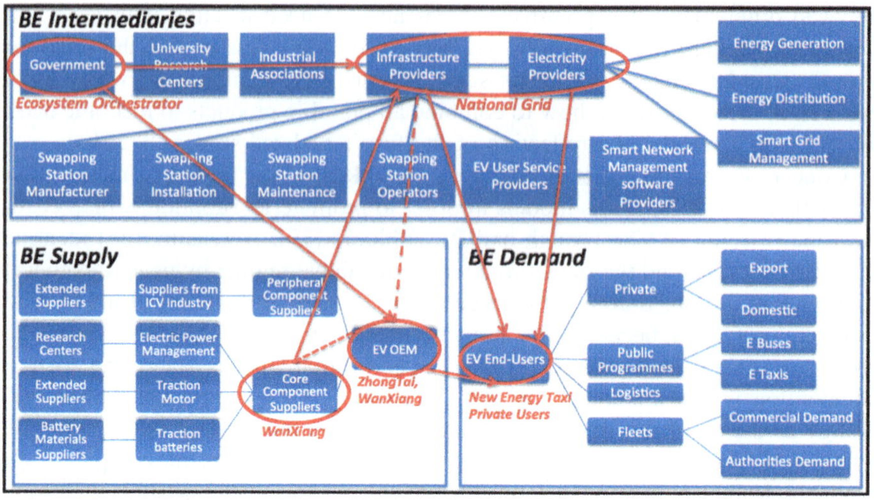

Fig. 3 Hangzhou's EV demonstration: battery swapping business model

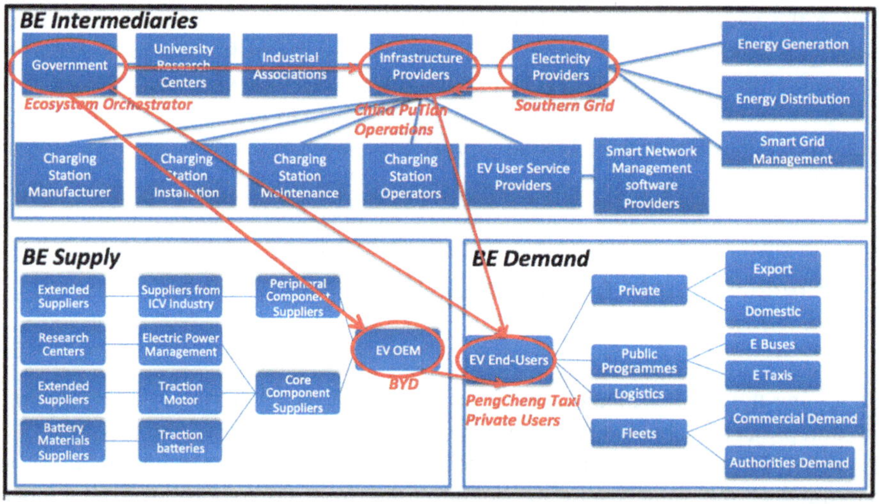

Fig. 4 Shenzhen's EV demonstration: battery-charging business model

In comparison, the EV ecosystem interaction illustrated in Fig. 4 depicts the battery-charging business model. It is apparent that while in both cases the local governments implementing the demonstration projects are acting as ecosystem orchestrators encouraging the development of EVs, BYD in the Shenzhen pilot project is operating in a traditional business model by selling EVs as a whole product compared with WanXiang and Zotye's Battery Swapping approach.

5 Findings

This paper has examined two contrasting business models in the context of the EV demonstration programme in China. While both battery charging and battery swapping models have encountered challenges, the pilot projects implemented by these two cities are considered to be the top ones within their business model domain among all the demonstration cities in China. In fact, officials and industrial players from other regions in China frequently visit both cities' EV pilot implementation offices seeking to learn from Hangzhou and Shenzhen; such that both cities perform experience and knowledge sharing activities on their EV operations. In the mean time, these co-existing business models are competing fiercely with one another. The logic behind these demonstration programmes is to implement trial operations at the very nascent stage of the industrial emergence with the main purpose of initial data collection as well as absorbing lessons from the feedback loop. The government therefore acts as ecosystem orchestrators coordinating and promoting different types of business models to be attempted and operated, hence, orchestrating ecosystem co-opetition. Consequently, the findings of this paper triggers potential practical implications for policymakers and industrial players who are tackling the challenges of stimulating a supportive business environment to promote the advancement of the EV industry in their respective countries.

Furthermore, the theoretical findings of this paper have revealed both the dynamic processes of business ecosystem development (Fig. 5) and static pictures of the ecosystem (Figs. 3 and 4). In particular, the research findings have shed more

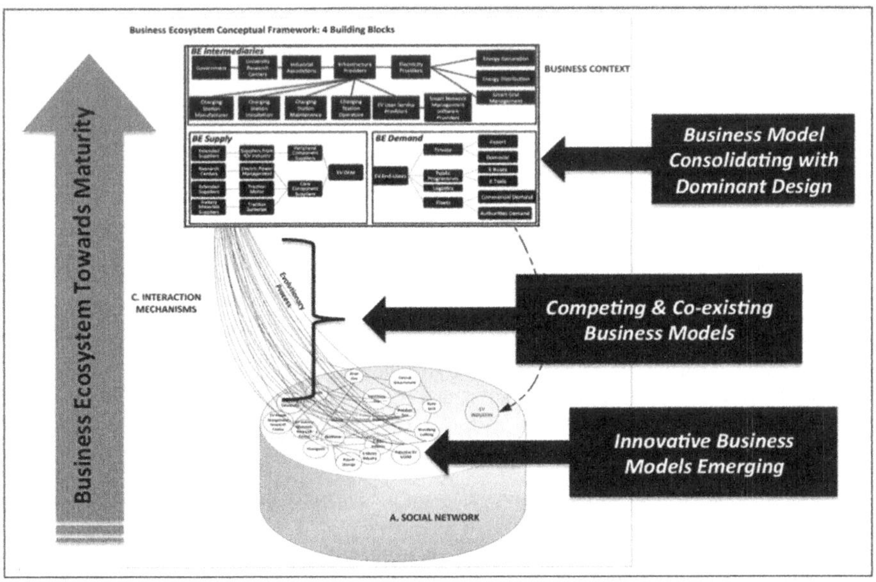

Fig. 5 Competing and co-existing business models in the context of ecosystem emergence

light on the roles of business models at the emerging stage of the ecosystem and their co-evolution (between business model and business ecosystem) along their development pathways (Fig. 5). From the embryonic development of the business ecosystem, innovative business models emerge at the very nascent stage; through the interaction mechanisms of the ecosystem players, competing and co-existing models are operated during the subsequent progressive stages of ecosystem growth until the consolidation of the business models with a dominant design.

6 Conclusion

In summary, this paper has provided insights concerning the development of the Chinese EV demonstration programmes in the cities of Hangzhou and Shenzhen through the conceptual lens of the business ecosystem framework. Currently, the government is acting as ecosystem orchestrators promoting different types of business models such that the Battery Swapping Model in Hangzhou and Battery Charging Model in Shenzhen are co-existing while competing fiercely; such actions of orchestrating ecosystem co-opetition allow policymakers and industrial players to contemplate alternative potential opportunities for stimulating a fertile environment in encouraging the progression of the EV sector. Theoretically, the business ecosystem concept has been defined and the relationship between the ecosystem and business models as well as their co-evolution has been captured. It is asserted that innovative business models emerge at the embryonic stage of the business ecosystem such that competing and co-existing models arises through the interaction mechanism between the ecosystem players; and the subsequent establishment of a mature business ecosystem allows the consolidation of the business models with a dominant design.

References

1. Moore J (1993) Predators and prey: a new ecology of competition. Harv Bus Rev 71:75
2. Moore J (1996) The death of competition. Harper Business, New York
3. Mannin B, Thorne C (2002) Demand driven: 6 steps to building an ecosystem of demand for your business. McGraw-Hill Companies, New York
4. Iansiti M, Levien R (2004) The keystone advantage: what the new dynamics of business ecosystems mean for strategy, innovation, and sustainability. Harvard Business Press, Boston
5. Quaadgras A (2005) Who joins the platform? The case on the RFID business ecosystem. In: Proceedings of the 38th annual Hawaii international conference on system sciences (HICSS'05), p 269b
6. Adner R (2006) Match your innovation strategy to your innovation ecosystem. Harv Bus Rev 84(4):98–107
7. Iansiti M, Richards GL (2006) Information technology ecosystem: structure, health, and performance. Antitrust Bull 51:77

8. Li YR (2009) The technological roadmap of Cisco's business ecosystem. Technovation 29 (5):379–386
9. Rong K (2011) Nurturing business ecosystem from firm perspective. PhD Dissertation, Department of Engineering, University of Cambridge
10. Shang T, Shi Y (2013) Rethinking the conceptual framework for business ecosystems: the evolutionary processes and key building blocks. European Academy of Management Conference, Istanbul
11. Long N, Waldemar A (2010) The electric vehicle sector in China: industry analysis, CEIBS (China Europe International Business School) Case Study, Code: CC-210-021
12. Amit R, Zott C (2001) Value creation in e-business. Strateg Manag J 22: 493–520
13. Dubosson-Torbay M, Osterwalder A, Pigneur Y (2002) E-business model design, classification, and measurements. Thunderbird Int Bus Rev 44(1):5–23
14. Hamel G (2000) Leading the revolution. Harvard Business School Press, Boston
15. Osterwalder A, Pigneur Y, Tucci C (2005) Clarifying business models: origins, present and future of the concept. Commun Assoc Inf Sci (CAIS) 16:1–25
16. Afuah A, Tucci C (2001) Internet business models and strategies: text and cases. McGraw-Hill, New York
17. Zott C, Amit R (2008) The fit between product market strategy and business model: implications for firm performance. Strateg Manag J 29:1–26
18. Johnson M, Christensen C, Kagermann H (2008) Reinventing your business model. Harv Bus Rev 86(12):50–59
19. Chesbrough HW, Rosenbloom R (2002) The role of the business model in capturing value from innovation: evidence from Xerox Corporation's technology spinoff companies. Ind Corp Change 11:533–534
20. Shafer SM, Smith HJ, Linder J (2005) The power of business models. Bus Horiz 48:199–207
21. Osterwalder A, Pigneur Y (2011) Business model generation: a handbook for visionaries, game changers, and challengers. Wiley, Hoboken
22. Casadesus-Masanell R, Ricart J (2010) From strategy to business models and to tactics. Long Range Plan 43:195–215
23. Björkdahl J (2009) Technology cross fertilization and the business model: the case of integrating ICTs in mechanical engineering products. Res Policy 38:1468–1477
24. Utterback J, Abernathy W (1975) A dynamic model of process and product innovation. Omega 3:639–656
25. Johann P, Koen F (2006) Toward a systematic framework for research on dominant designs, technological innovations, and industrial change. Res Policy 35:925–952
26. Raji S, Lilien GL, Arvind R (2006) The emergence of dominant designs. J Mark 70:1−17
27. Pek-Hooi S (2010) Network patterns and competitive advantage before the emergence of a dominant design. Strateg Manag J 31(4):438–461

EVs to Reduce Dependence on Imported Oil: Challenges and Lessons from Maui

Anne Ku

Abstract Hawaii's geographic isolation and historical dependence on imported fossil fuels are the primary cause of its residents having to pay the highest energy prices in the USA. To reduce oil dependence in transportation, the State of Hawaii introduced EV-friendly policies (in 2009) and financial incentives (in 2010) for an early adoption of plug-in electric vehicles (EV) and deployment of associated charging infrastructure. In 2011, University of Hawaii Maui College led a consortium of grant and cost-share partners to plan for mass EV deployment for Maui County in a 2-year project called "Maui Electric Vehicle Alliance." The "planning" involved regular meetings and discussions among stakeholders, continuous outreach and education, and learning through implementation. An organized group of stakeholders acting as a central repository of information and coordinator of EV-related events while providing opportunities to educate and engage the community is essential to building confidence in this new EV technology and cultivating a change of driver attitude and behavior.

Keywords Big Island · DBEDT · Charging stations · Kauai · Lanai · Maui · Molokai · Oahu · Plug-in electric vehicles · Battery storage · CHAdeMO · Chevy VOLT · DC fast charger · Electric vehicles · EV · EVSE · EV readiness · Feed-in tariff · Hawaii · Hawaii Island · Hawaii state energy office · HCC · HCEI · HEPF · Honolulu · Honolulu clean cities · JUMPSmartMaui project · Maui College · Maui EVA · Mitsubishi iMiev · NEM · Net energy metering · Nissan LEAF · PHEV · Photovoltaic · Range anxiety · Renewable energy · Smart grid · Solar · Tax rebate · UHMC · University of Hawaii Maui College · Vehicle to grid · Vehicle to home · V2G · V2H · Wind · Wind power

A. Ku (✉)
University of Hawaii Maui College, 310 West Ka'ahumanu Avenue, Kahului, HI 96732, USA
e-mail: anneku@hawaii.edu

D. Beeton and G. Meyer (eds.), *Electric Vehicle Business Models*,
Lecture Notes in Mobility, DOI 10.1007/978-3-319-12244-1_14

1 Why Hawaii Needs EVs

On the surface, the Islands of Hawaii [1] have the perfect set of push and pull factors to welcome plug-in electric vehicles (EV) [2].

High gasoline prices push for the consideration of alternative fuels for transportation. The Islands' abundant renewable energy potential, high rooftop solar penetration, excess wind power at night, limited driving distances, and sustainability-minded residents all provide ideal conditions for electric vehicle adoption. At the same time, high labour and electricity costs and a conservative mindset impede a more rapid development of EV charging infrastructure.

Hawaii's traditional dependence on oil for ground, sea, and air transportation makes it an oil-based economy prone to oil price fluctuations. Oil for electricity is refined from the same barrel imported for transportation purposes and constitutes 90 % of the energy used. A tenth of Hawaii's gross domestic product is spent on energy, most of it for imported crude oil and petroleum products. As a result, Hawaii is the most petroleum-dependent state in the U.S.

As the most isolated population centre on earth, Hawaii's remoteness and vast distances from production regions result in high shipping and labour costs as well as sensitivity to supply shocks and price spikes. Each inhabited Island has its own isolated grid, forcing a need to maintain high reserves for load swings. Hawaii's residential and commercial electricity prices (as shown in Table 1) are more than three times the national average, and more than double its nearest competitor, the State of New York (see Fig. 1).

While Hawaii's abundant renewable energy potential promises a viable alternative to fossil fuel-dependence, harnessing such intermittent sources of energy requires grid-integration capability, infrastructure upgrades and a utility business model that supports large-scale distributed generation. The battery storage capabilities of plug-in electric vehicles and vehicle-to-grid and vehicle-to-home technologies may hold the key to this energy transformation.

To reduce Hawaii's dependence on imported oil, the State of Hawaii began a unique partnership with the U.S. Department of Energy (DOE) in setting up the Hawaii Clean Energy Initiative (HCEI) in 2008. The HCEI goal in transportation is to reduce petroleum consumption by 70 % or displace 385 million gallons of petroleum by year 2030 [4].

Reducing petroleum use in transportation is fraught with challenges. Unlike the electric power sector, fuel use in transportation is not regulated or provided by utilities. Vehicle efficiency standards are set federally and not subject to state authority.

Table 1 Average retail electricity prices (total revenue/total sales) September 2013 [3]

US cents/kWh	Molokai	Lanai	Kauai	Hawaii	Maui	Oahu	State
Electricity retail prices, all sectors	46.00	42.62	42.60	39.44	36.44	29.73	32.22
Residential	47.49	47.66	44.20	41.61	37.71	33.80	36.19
Commercial	45.33	40.75	41.56	38.20	35.82	28.43	30.73
Street lights	44.73	46.62	58.56	40.31	35.28	31.08	33.97

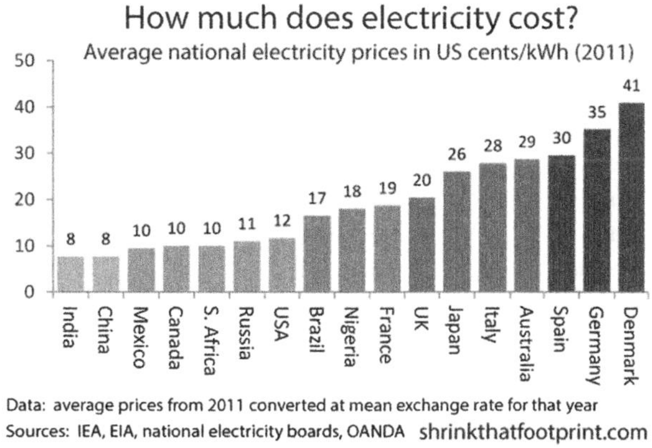

Fig. 1 Hawaii's electricity prices are on par with the most expensive in the world

The substantial amount of fuel used for military and commercial air transportation is beyond the control of Hawaii State policies and laws. No state agency has the mandate, funding and jurisdiction necessary to effectively promote transportation fuel reduction initiatives [5].

To meet the HCEI goal in transportation, the State of Hawaii has committed to a comprehensive transportation strategy that includes the adoption and integration of EVs and charging networks. These targets effectively translate to 10,000 plug-in electric vehicles (EVs) by 2015 and 40,000 by the end of 2020. At end of March 2014, 2,375 EVs have been registered in the State of Hawaii [3].

Push factors aside, the geography and climate of the Hawaiian Islands provide ideal conditions for electric vehicle deployment. Range anxiety is kept in check, as one Colorado native puts it, "There's no danger of driving too far into another state and run out of charge." Elevation is a consideration, however.

To see how EVs can complete the picture, it helps to understand the geography and economy of the Hawaiian Islands.

2 Hawaii: The State, The Counties, The Islands

The Hawaiian Archipelago spans some 100 Islands, spreading over 1,500 miles in the Pacific Ocean, by far the widest state in the United States and the only one comprised entirely of Islands, as can be seen in Table 2 and Fig. 2. The State of Hawaii is divided into four counties over one uninhabited and seven populated Islands, namely:

- Kauai County comprises Kauai Island and Niihau.
- Honolulu County, officially the City and County of Honolulu, comprises Oahu and the small Islands northwest of Kauai and Niihau extending from Nihoa to Kure except for Midway. Honolulu is the only city in the state. To avoid confusion, Oahu is used in subsequent tables to represent this county.
- Maui County comprises Kahoolawe, Lanai, Maui Island, and Molokai.
- Hawaii County comprises Hawaii Island, a.k.a. the Big Island.

Table 2 The Islands of Hawaii

		Land area		Highest point
Island (from west to east)	Also known as	Square miles	Square kilometers	Feet
Niihau	Forbidden Isle	70	180	1,250
Kauai	Garden Isle	552	1,4301	5,243
Oahu	Gathering Place	597	1,545	4,003
Molokai	Friendly Isle	260	637	4,951
Lanai	Pineapple Isle	141	364	3,366
Kahoolawe	Target Isle	45	116	1,483
Maui	Valley Isle	727	1,883	10,023
Hawaii	Big Island	4,028	10,433	13,796

Fig. 2 Map of Hawaii

Farthest west is the privately owned Island of Niihau, the smallest of the inhabited Islands, 18 miles from the coast of Kauai. Farthest east is the youngest and largest Island of Hawaii, also known as Big Island, famous for its five volcanoes.

The Island of Oahu is home to 70 % of the state's 1.4 million resident population (as illustrated in Fig. 3). Maui County includes Maui Island, Molokai, Lanai, and the uninhabited Island of Kahoolawe.

Some eight million visitors travel to the State of Hawaii every year. While Fig. 4 shows the geographic distribution within Hawaii, Fig. 5 additionally displays the

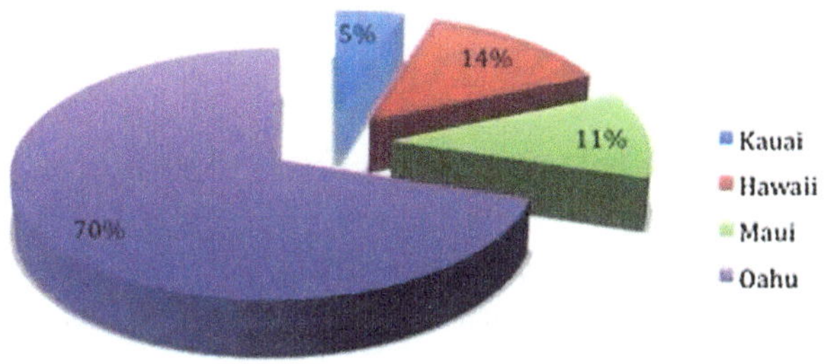

Fig. 3 Resident population total 1.4 million in the state [6]

Fig. 4 Visitor population, average daily visitor census, arrivals by air [6, 7]

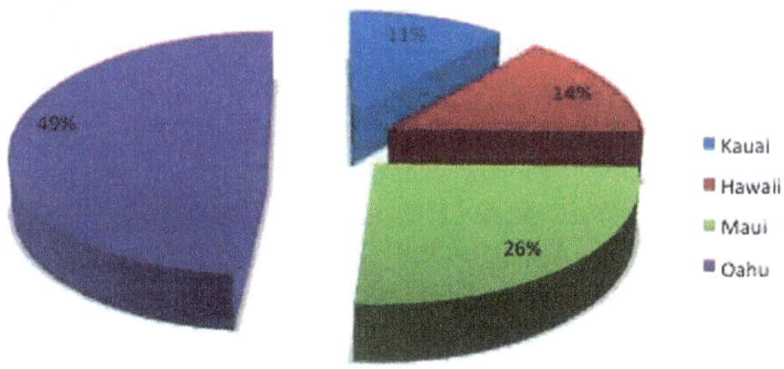

Fig. 5 Average daily visitors a noticeable presence (vs. resident population) by county [6]

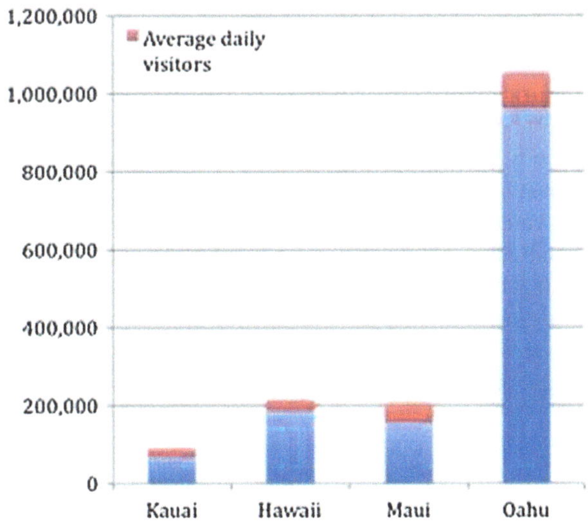

number of visitors in comparison to residents. Tourism is the largest single contributor to the state's gross domestic product, representing about 21 % of its entire economy. In 2012, its best year on record, daily visitor spending topped $39 million; state tax revenue reached $1.58 billion; and the industry supported 167,000 jobs. That year, visitors spent more than $14 billion in Hawaii [7].

As a labour-intensive industry, tourism has generated more jobs than any other sector of the state economy, accounting for over one-third of all jobs in Maui (37.9 %) and Kauai (36.6 %), followed by Hawaii (20.6 %) and Honolulu (12.0 %) [8].

3 State Policies and Incentives for Early Adopters

To meet the Hawaii Clean Energy Initiative goals, the State of Hawaii encouraged early investment of EVs and associated infrastructure through law-making and financial incentives.

In 2009, the Hawaii State Legislature passed several innovative EV-related bills [9]. After amendments in 2012, the major ones require certain properties to install EV-designated parking stalls and charging access, State and County to allow free parking for EVs, State and County agencies to follow a hierarchy of alternative-fuel technology preferences when procuring or leasing light-duty vehicles, and that residents in multi-unit dwellings cannot be denied the right to install charging equipment.

To kickstart mass EV adoption and infrastructure deployment in 2010, the Hawaii State Energy Office through the Department of Business, Economic Development and Tourism (DBEDT) administered $4.5 million from the American Recovery and Reinvestment Act (ARRA) stimulus funds for tax rebates and financial awards to motivate early adoption of EVs and deployment of charging stations. The progress of the Hawaii State EV Ready Rebate Program Fund Uptake, as an illustrative example, can be seen in Fig. 6. Figure 7 shows how charger and EV rebates were distributed among the Islands.

$1.9 million was given to owners of plug-in vehicles (455) and charging stations (279) between August 2010 and April 2012. These paid for up to 20 % of purchase price with maximum of $4,500 per plug-in electric vehicle and up to 30 % of cost of charging equipment and installation with a maximum of $500 per system.

Another $2.6 million was awarded to six organizations for the systematic installation of electric vehicle chargers across the State; public education and outreach including an EV Ready Guidebook [12] (for the installation of commercial charging stations); introduction of EVs to rental car and county fleets; car-sharing services within the hospitality industry; and an online permitting system for charger installations at single-family residences on Oahu.

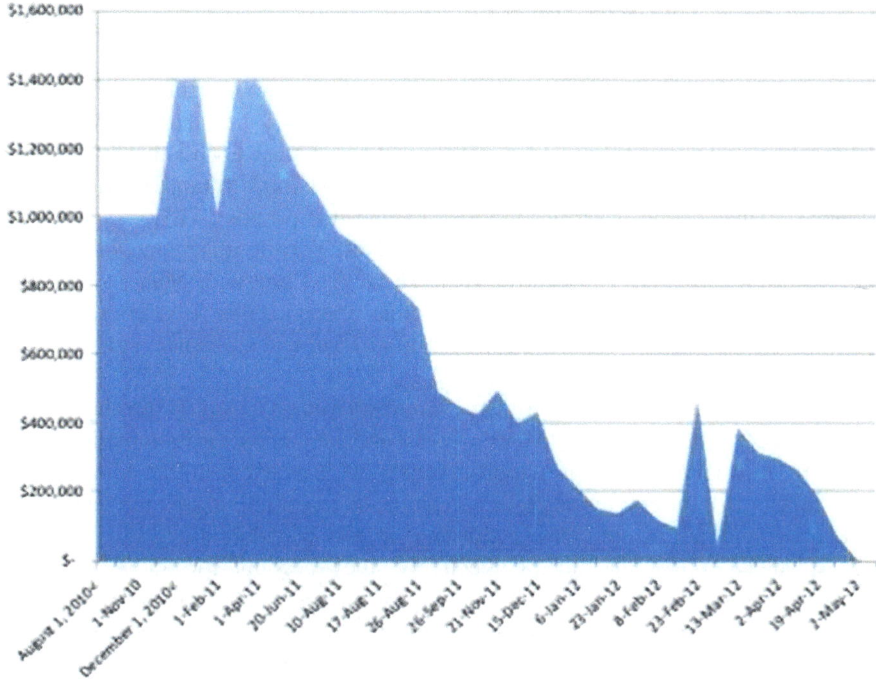

Fig. 6 Hawaii State EV Ready Rebate Program Fund Uptake [10]

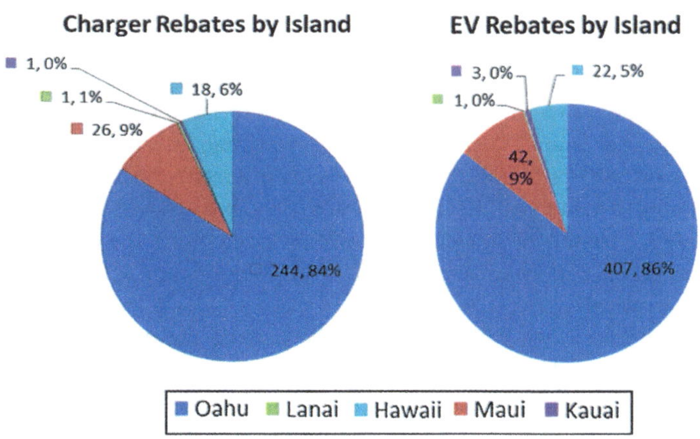

Fig. 7 Tax rebates for charging stations and electric vehicles [11]

4 Maui as a Test-Bed for EVs

In June 2011, University of Hawaii Maui College, in a consortium with partners Hawaii State Energy Office/DBEDT, Honolulu Clean Cities, and University of California San Diego, submitted a proposal that won an EV-readiness planning grant from the Department of Energy for the project called "Maui Electric Vehicle Alliance." This section describes the idea behind this proposal [13].

In 2013, Maui earned the top spot in Condé Nast Traveler magazine's list for top Islands for the twentieth consecutive year. The review boasts of Maui as an Island with an "abundance of activities" and the online version states "Readers rave about this veritable paradise, calling it a combination of tropical ambiance and American comforts."

More than two million people visit Maui each year. On any given day, one in three people are non-residents. Maui's visitor industry is by far its most important economic sector, for it generates approximately 80 % of every dollar. Most visitors on Maui stay in condominiums, timeshare units, or resort hotels, and the majority of these accommodations are densely clustered in two corridors along the south and western areas of the coastline. Collectively, they account for 85 % of visitor accommodations.

The Island of Maui is approximately 48 miles long and 26 miles wide, totaling 727 square miles. With the exception of the long and winding scenic "Road to Hana," the distances between common destinations are short, often 30 miles or less. This creates an appropriate environment for EVs, even at a time when "range anxiety" is still a strongly perceived barrier for consumers elsewhere.

On average, 80 % of the visitors choose to travel primarily by rental car. Local rental car companies estimate that over 15 % of all passenger vehicles on Maui are rental cars, a percentage that may be unmatched anywhere else in the United States. Moreover, passenger vehicles make up over 75 % of the total number of registered motor vehicles on Maui. The remainder is small and medium-sized commercial trucks and vans, and public transportation vehicle fleets—reflective of an economy that is strongly dependent on services rather than heavy industry. This high proportion of light, passenger vehicles means that accelerating the transition of rental car fleets to EVs will have an immediate and significant impact on petroleum consumption overall.

Driving rented electric vehicles would be part of the experience of paradise. After an extended test drive for the duration of their stay, visitors would return home with the added knowledge and confidence to consider buying an electric vehicle of their own. Visitors would also help spread the word about EVs and their driving and charging experience on Maui.

Strategic use of the rental car and resort hotel model may solve the chicken or egg dilemma of what comes first—EVs or the charging infrastructure. It also addresses Hawaii's growing concern of how to bridge the chasm between early adopters and mass market electrification of transportation. One way to test the idea is to try it on a smaller scale. If this works, the model could be replicated to other Islands.

Until the range of electric vehicles advances to that of gasoline-powered vehicles, EV drivers will need to be assured of ample opportunities to stop and charge, without having to experience long wait times or driving elsewhere because access is unavailable due to inoperability or long queues. To eliminate range anxiety, a complete charging infrastructure needs to be installed and access made available at hotels, condos, and other visitor accommodations as well as shopping malls and tourist attractions.

The transition will not be complete, however, until the source of the electrification is also transformed.

Maui has numerous renewable energy options (as shown in Fig. 8), including sun, wind, sea, and land, and potential for geothermal, hydropower, and biomass resources. It is served by the Maui Electric Company (MECO), a regulated (investor-owned) utility, which has about 260 MW of generation capacity with a peak of 190 MW. It leads the nation in installed PV per capita [14]. The excess or curtailed energy from Maui's three wind farms could power Maui's passenger vehicle fleet overnight.

Like other Islands in the Hawaiian chain, Maui's isolated grid means that gaps cannot be filled by buying energy from other utilities or selling excess energy when capacity is exceeded. However, this isolation also makes Maui an excellent "laboratory" for smart grid testing and analysis of EV impact, a factor not lost on energy technology companies.

Taken together, Maui's high profile visitor industry, the reliance on rental car industry vehicles, the clustering of rental vehicles in tourist destinations and hotels, the short driving distances, and Maui's selection as a demonstration site for smart grid technologies—all create an ideal environment for transitioning the rental car fleet to one containing a high percentage of EVs with a concurrent build out of public charging stations.

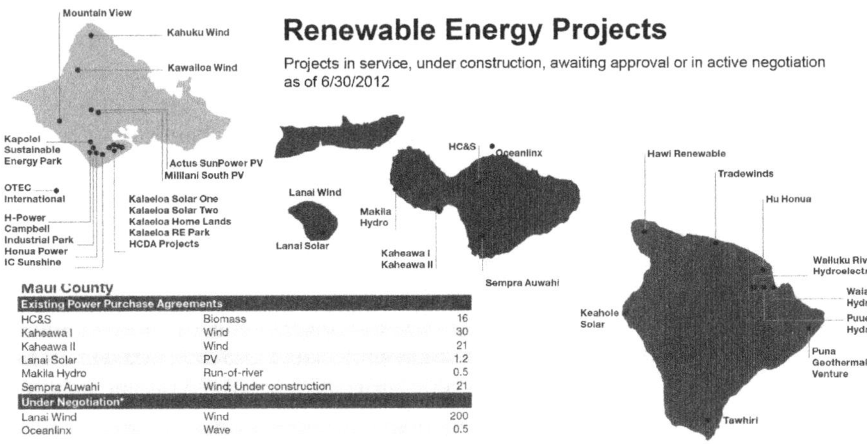

Fig. 8 Renewable energy projects in Maui county [15]

Hypothetically, targeting rental car fleets will allow Maui to achieve the highest penetration of EVs per capita in a relatively short time frame. Building out public charging infrastructure not only supports the visitor industry but also will influence resident drivers to purchase EVs, as they observe a ready supply of public charging stations in use by visitors.

Maui's renewable energy projects, smart grid programs, and Island geography make it an ideal candidate for EV showcasing.

5 Maui Electric Vehicle Alliance

Initially awarded for 1 year from October 1st, 2011, the Maui EVA project was extended twice, finally ending on January 31, 2014. The extension allowed a critical mass of EV adopters and forward momentum to be built. Over this period, the role of Maui EVA evolved and grew in importance.

As an open access alliance, it served as an approachable organization to engage and create dialogue regarding the EV market on Maui and other neighbour Islands, organizing different stakeholders to meet and discuss issues to do with EV readiness, building confidence and trust in an unknown technology, raising awareness and educating the community through a multitude of widely-publicized events (working group meetings, EV conferences, seminars, meet-ups, test-drives), local media (easy-to-read educational articles in own newspaper column, regular press releases, 14 one-hour TV episodes broadcasted weekly and available on-demand), social media (Facebook, Twitter, Linked-In, YouTube), and publication of key reports [16], serving as a central point of contact for EV expertise and a central repository of best practices and updates.

Maui EVA's strategy was to create a big splash by including the local press and actively engaging the community from the outset, getting national and international attention to spotlight on Maui, and attracting EV experts to come to Maui. Shortly after kick-off, the project was named the third most innovative by Green Tech Media [17]. In summer 2013, Maui joined the World EV Cities project and website [18]. At every opportunity, the project welcomed new participants to the EV conversation, thereby enlarging the outreach and ensuring inclusiveness.

By the end of the project in January 2014, Maui had achieved the highest EV registration per capita in the state (Fig. 9 and Table 3), with the state vying with Washington for the top position in % EV sales per new car sales in the USA [19]. Already in September 2013, Navigant Research predicted that Hawaii will lead the nation in EV sales per new car sales by 2022 [20].

Coincident to the launch of Maui EVA was the signing of the $37 million Japan–US Island Smart Grid Demonstration project [21], later rebranded JUMP-SmartMaui [22]. Representatives from this project joined the working groups of Maui EVA and requested assistance in achieving the target of 200 Nissan LEAF owners as EV volunteers (Nissan LEAF registrations shown in Fig. 10).

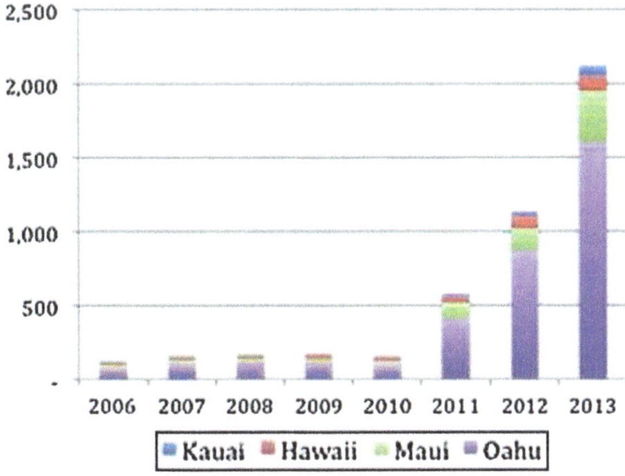

Fig. 9 Cumulative EV registrations—Maui gains momentum [3]

Table 3 Maui has the highest EV per capita comparisons across the counties in the state

County	EVs	Residents (Res)	Average daily visitors (Vis)	EV/Res	EV/(Res + Vis)
Kauai	70	67,701	20,814	0.00103	0.00079
Hawaii	108	186,738	26,550	0.00058	0.00051
Maui	383	156,764	49,481	**0.00244**	**0.00186**
Oahu	1,661	963,607	88,979	0.00172	0.00158
Total (state)	2,222	1,374,810	185,824	0.00162	0.00142

EV registrations as at end January 2014 [3, 6]

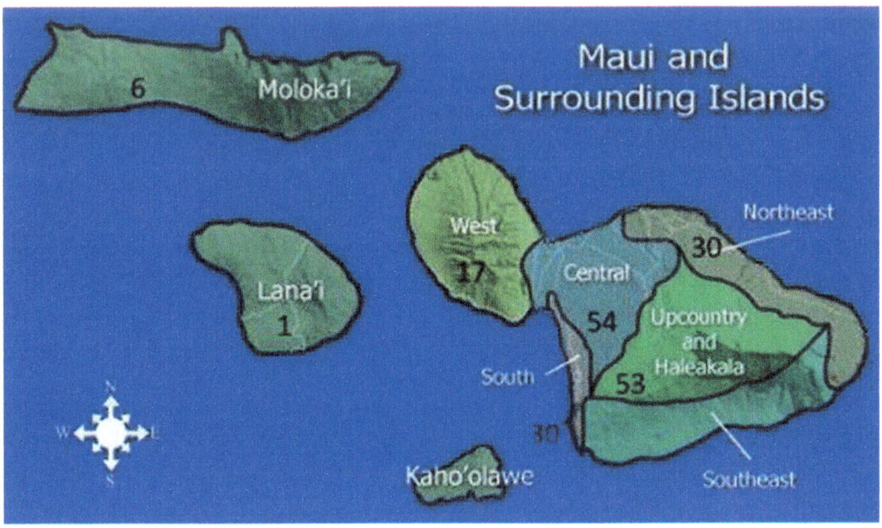

Fig. 10 Nissan LEAF registrations in Maui County @ October 2013. *Source* POLK Database (191 LEAFs out of 304 EVs registered by end Oct 2013)

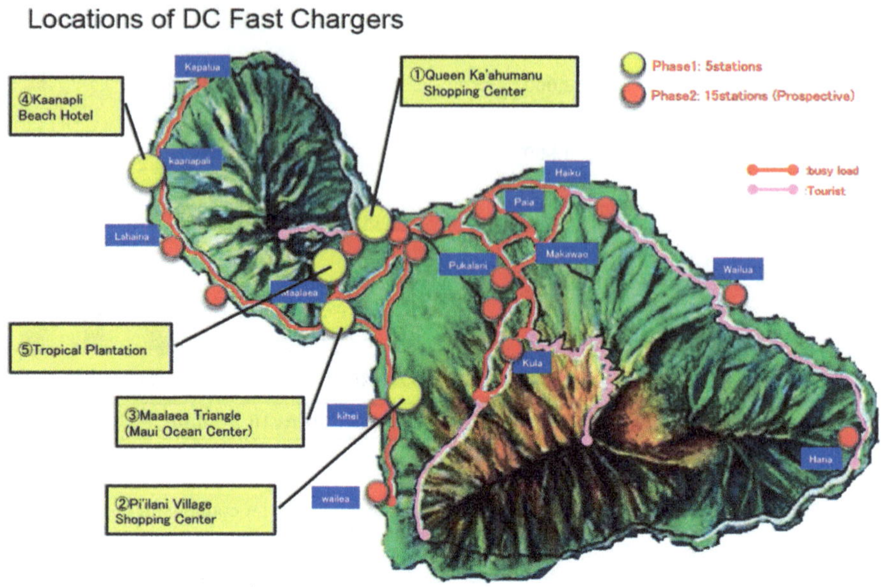

Fig. 11 Locations of Hitachi DC fast chargers based on traffic and demand [22]

While the funding for the Maui EVA project allowed no budget for equipment or infrastructure, the JUMPSmartMaui project is focused almost entirely on equipment and infrastructure—testing and data collection on smart grid technology, battery storage, electric vehicles, and DC fast chargers—and knowledge transfer between Japan and Hawaii, specifically Okinawa and Maui. Hitachi installed its quad-port DC fast chargers at five locations on Maui, with a view to expand further in phase two (as shown in Fig. 11).

6 Residents to Prepare for Visitors

Half-way through the Maui EVA project in April 2012, the project's Visitors Industry Working Group declared that the initial vision of "EVs in Paradise" must be tried and tested on local residents first, before it would be prepared to market this novel idea. The group was concerned about the lack of vehicle models with ample luggage space and sufficient range for the needs of the typical tourist.

The Visitor Industry Working Group reported that hotels, condos, and other property owners were hesitant to invest in charging stations, unless mandated by their hotel brands, e.g. Marriott, Westin, etc. Having recently emerged from the recession of 2008/2009 that caused widespread cutbacks, layoffs and bankruptcies, these businesses are wary of setting aside time and resources on what's considered a significant investment and risk before "the EVs show up." As energy is the next highest expense after personnel, most large properties have indicated they could subsume this cost into the "resort fee" paid by their guests.

Without a complete and comprehensive charging infrastructure in place, rental car companies are reluctant to provide electric vehicles that cannot be conveniently recharged. One way to resolve this "chicken or the egg" problem is to have rental car companies partner with hotels to create a demand for EV rentals. Another way to create demand for EVs is for employers to make it possible for their staff to rent EVs on business trips. Currently, only Oahu and Maui Islands have car rental companies offering EVs for rent.

Whether resident or visitor, there are three kinds of customers inclined to rent EVs. Those visitors that already own or use a particular kind of EV at home may rent EVs for the familiarity of it. Those residents or visitors who have never driven an EV but are curious about the technology may rent for an extended test-drive. Residents are especially encouraged to do so while their regular car is being fixed, i. e. rent a replacement car that is an EV. Residents may also choose an EV on business if it is allowable and paid by their employers. Finally, rental customers who are price-sensitive would choose an EV if it's less expensive than their usual preference. If EVs are the only vehicles available in the inventory and customers must rent a vehicle to get around, then price would be no object (Table 4).

Table 4 EV driving and charging patterns of residents versus visitors

	Residents	Visitors
Driving routes and patterns	Predictable, routine, familiarity with geography (distance and elevation), optimized, short-cuts, knowledge of alternative or back roads	Unfamiliar with local routes and language (names of streets and towns are in Hawaiian—not easy to remember or pronounce for the typical visitor). Unpredictable, not optimised, spontaneous, unfamiliar, highly variable, wrong turns, may drive longer and less efficiently than residents, more frequent stops, U-turns, longer routes. Prone to getting lost
Charging station locations	Home, work, known locations	Hotels, tourist destinations, shopping malls
Back-up plans	Call other EV owners or known acquaintances or friends	Ask strangers or call hotel or travel agency
Charging patterns	Routine and predictable	Sporatic, may stop to charge more frequently than necessary due to anxiety, lack of familiarity, and poor judgment, risk of under-estimating distance and elevation
Types of vehicles preferred	Sedan, hatchback, pick-up trucks, SUVs, four-wheel drives, off-road vehicles	Convertibles, SUVs, cars with big trunk space for luggage, sedans. Four-wheel drives, off-road vehicles

Table 5 Okinawa versus Maui

Island	Land mass square miles (km^2)	Resident population	Annual visitors	Landscape	EVs
Okinawa	463.7 (1,201)	1.3 million	5 million	Urban	Nissan LEAFs
Maui	727 (1,883)	158,000	2 million	Rural	Nissan LEAFs, Chevy VOLTS

The idea of EV rentals for tourists is not a new one. Orlando, Florida and Las Vegas, Nevada are targets for EV rentals. Okinawa, Japan has an extensive charging network in place for the 220 EVs introduced in February 2011 [23]. Table 5 gives an overview of the differences between the Islands Okinawa and Maui.

The first 3 years of Okinawa's EV rental and DC Fast Charger experience showed a disappointingly low rental car utilization rate, lower than expected sales of second-hand EVs, and a lack of information sharing between the stakeholders. The Okinawa case study [24] highlights a dire need to build confidence in the new technology, particularly for intermediaries and gate-keepers who influence the final decision of customers who may rent or purchase EVs.

7 Lessons Learned from Maui EVA

In many ways, the old adage "it takes a village" is relevant to the Hawaii example insofar as a collective contribution of stakeholders is needed to transform an oil-based economy through EVs at every level.

Besides the Federal and State government's leadership in providing financial and policy incentives, the electric utilities in Hawaii are among the 6 % in the USA that have introduced time of use (TOU) rates for EVs [25]. In 2013, two new tariffs were passed to allow utilities to install DC Fast Chargers and charge a TOU rate for their use [26].

Two rental car companies in Hawaii continue to offer the Nissan LEAF and Chevy VOLT for rent on Oahu and Maui, including special rates for local residents. A third, GreenCar Hawaii ceased its operations in Kauai and Oahu in late 2013 after its acquisition by JustShareIt.

The upfront $1 K cash savings and no haggle leasing price of Nissan's Vehicle Purchase Plan (VPP) available to employees of all accredited universities and colleges in the USA caused a surge of new LEAF owners—more than 25 within a 6 month period at UH Maui College alone.

To meet its target of 200 LEAF volunteers and 40 home volunteers, JUMP-SmartMaui offered a cash bonus for certain kinds of EVs, free residential charging equipment, and free EV charging at Hitachi's DC fast charge locations on Maui. At time of writing, Maui is the only place in the USA installed with Hitachi DC fast chargers.

EV owners themselves are influencing EV sales through the so-called "coconut wireless." The top Nissan LEAF salesman on Maui reports that "for every EV sold, five more people come to enquire. These are the family, friends, neighbours and colleagues of the customer." He adds, "They already know what they want when they come to the shop. I don't do the selling, just the administration." Word-of-mouth is a powerful instrument of communication, particularly in Island communities where the degree of separation is one not six. Business is done by who you know, not the yellow pages. Early adopters who are well-known, respected, and vocal have much to contribute to the continued adoption of EVs.

Oahu's rapid initial growth of EVs can be attributed to proactive EV dealerships, influential early adopters, favourable state tax and policy incentives (free HOV lanes and parking). Meanwhile, Hawaii Island's slow and stagnant EV growth can be attributed to the complete absence of DC Fast Chargers and Nissan LEAF dealerships. Kauai's slow progress can be attributed to the absence of an organized group of stakeholders that champion EVs.

The strategic placement of a single DC Fast Charger provided with free 24/7 access can make a huge difference in EV adoption. Consider Maui and Kauai. The very public announcement of the first DC Fast Charger installed at the centrally-located County Building parking area in Wailuku in September 2012, together with the National Plug-in Day "ride and drive event" at UH Maui College campus the next day, boosted new EV sales on Maui. A similar ARRA-funded AeroVironment DC Fast Charging Station and Level 2 Charger installed at a private hotel in Kauai did not have the same effect. EV growth in Kauai has been far slower than in Maui.

7.1 Planning and Coordination to Avoid Inconvenience

The demand for EVs has outstripped supply and infrastructure on several occasions. When Better Place Hawaii closed business in March 2013, a year after showcasing their ARRA-funded Level 2 charging stations in seven locations on Maui in the widely publicized "Drive Electric Maui" event, the director retained 100 keyfobs on the presumption that the demise of the company would dampen EV adoption. Contrary to his expectations, new EV owners soon depleted his supply of fobs. Without these keyfobs, new EV owners were unable to access these free charging stations.

Other instances of EV adoption outperforming expectations include the early depletion of EV license plates in June 2013. Vehicles with EV license plates get free parking privileges at state and county lots, including airports.

If the pace of EV adoption continues at present rates, there will be a real risk of shortage of qualified and certified EV mechanics. At present, there is only one certified Nissan LEAF mechanic at the Maui dealership serving four Islands: Maui, Molokai, Lanai, and Hawaii Island, and one qualified VOLT mechanic serving Maui, Molokai, and Lanai. Although EVs are currently under warranty, there's still a concern of having to wait for service if there is a problem and what warranty does not cover (e.g. accidents). If there are any recalls, there will be a massive backlog of getting the EVs serviced and the recall issue addressed. For instance, a new LEAF owner in Molokai chose to pay out-of pocket to ship his vehicle to a dealership on Oahu for warranty-covered service instead of waiting more than two weeks for Maui's technician to become available for him.

7.2 The PV to EV Link: The Key to Greater EV Adoption

At first glance, it would seem that Hawaii's high electricity prices discourage EV adoption. Though much less volatile than gasoline, electricity prices have not decreased over time and are unlikely to do so. Unless EV drivers can get free or subsidized charging outside the home, there is a perception that fueling EVs is not cheaper than gasoline. If anything, it's less convenient because the shorter range of EVs requires greater frequency to charge. At residential rates of 40–45 cents/kWh, a Nissan LEAF would require $10 to charge from completely empty to full.

Homeowners, who have installed sufficient photovoltaic panels to include EV charging, have the benefit of capping their monthly electricity bill at $18 through Net Energy Metering (NEM) or Feed In Tariff (FIT) arrangements with the local utility. However, energy credits accumulated during a calendar year cannot get transferred to the next year or to another electricity customer. One Maui resident said that he would rather use it up than forfeit the credits to the local utility. Instead of cranking up the air conditioning which is not needed towards the end of the year, he bought a Nissan LEAF. For part-time residents, also known as "snow birds," the combination of PV on roof and EV in garage is a perfect symbiosis: accumulate credits during the summer when they are off Island and use it to charge their EVs when they return in the winter.

With Hawaii leading the nation in terms of percentage of utility customers with photovoltaic systems, it is very likely that the correlation between PV and EV ownership is higher than the 39 % found in California [27].

The PV to EV link is potentially much higher on Maui where the percentage of single family dwellings is higher than on Oahu and California. At time of writing, there is a backlog of requests for new roof-top solar installations. Citing distribution grids as over-saturated, risking voltage spikes, damage of appliances, electronics, and utility's equipment, the local utilities require time to conduct interconnection studies. This is the big elephant in the room—many frustrated home-owners and their solar installers having to pay for such studies and wait indefinitely for a connection. Once this barrier is lifted, another floodgate for EV adoption will be opened.

7.3 Incentives and Penalties for Charging Infrastructure Deployment

Hawaii is one of the few states that require certain types of properties to install charging stations and multi-unit dwellings to oblige to residents' requests for charging access. Act 089 [Senate Bill 2747 SD1 HD2] [28] requires owners of properties with 100 or more parking spaces for "public accommodation" to make available at least one designated parking stall exclusively for EVs, and with access to charging. As few properties have complied since its introduction, this law is being amended in the 2014 state legislative session to include an escalating penalty for noncompliance.

Installing charging stations to comply with the law has caused some property owners to consider providing EV charging and parking a chore rather than a business opportunity. While some hotels have installed such facilities as an added amenity to avoid turning away guests with EVs, the majority prefers a wait-and-see attitude and learn from those who have installed.

Workplace charging is slowly catching on. University of Hawaii Maui College timed its four duo-port level two charging stations with the construction of its long-awaited solar carport, a project to reduce the college's electricity bill. Twenty standard three-prong outdoor sockets are distributed under the carport to provide level one charging access for all types of EVs including neighbourhood electric vehicles, electric bicycles, and electric motorcycles. This is an example of the largest investment in workplace charging (to-date) on Maui.

In general, charging stations cost more to install in Hawaii compared to the rest of the nation because of high cost of import and labour. Unique site considerations can also result in delays and additional expense. These include Special Management Areas (SMA), permits, trenching, upgrades, environmental factors, flood zones, and security. Based on data collected by the Hawaii State Energy Office, charging station installations in Hawaii can range from $4,000 to $25,000, with a relatively simple project in Hawaii typically costing approximately $6,000–$8,000 per station. Hawaii is also a challenge for installation of charging stations because of flood control mandates and SMAs that form the coastal zone management system. Most of the properties frequented by tourists are located in these areas, and under some conditions the charging stations must be elevated or are subject to additional requirements.

Environmental conditions on a tropical Island also pose a challenge for installation and maintenance. High salt content in the air, humidity, and full sun exposure can lead to equipment degradation and malfunction. For example, staff at one Maui installation site reported that the equipment rusted within months of installation. A level two charging station, located under direct sun, soon stopped working less than a year after installation. Owners therefore need to purchase charging equipment with protective covers or anticipate higher replacement costs because of shorter lifespans for the equipment.

The developing EV market in Hawaii allows participants to research and gain insight into EV infrastructure opportunities and challenges. The Honolulu Clean Cities report [29], the joint Berkeley and Maui College policy paper [30] and two other aforementioned reports from this grant share best practice recommendations and lessons learned, all of which are linked from the Maui EVA project website at http://www.mauieva.org.

Through the Maui EVA project, three stages of market share growth were identified. In the short term (2011–2013), stakeholders were invited and organized to examine and discuss the challenges and opportunities. Lessons learned from early adopters were communicated. Extensive public awareness was made and technical reports were translated to layman's language. In the medium term (2013–2015), EV-related policies are further developed, including those for workplace charging. New EV models enter the market. Workforce training for EV-certified mechanics and first responders become critical. In the long term (2016–2020), once a significant number of residents have adopted EVs and all major tourist destinations have charging access, the visitor industry gets ready to introduce EVs to tourists [31]. Meanwhile, the need for information, access to experts, policy development, and public relations will continue.

In many respects, the challenges and opportunities for EVs in Hawaii have more in common with other Islands in the world than the other 49 states of the USA.

References

1. To avoid confusion, the ʻokina and kahakō used in the Hawaiian language are intentionally not used here. For instance, Hawai'i is spelled Hawaii. O'ahu is spelled Oahu. Hāna is spelled Hana
2. For simplicity purposes, EVs in this paper refer all vehicles that have a plug to an electricity source. At time of writing, Tesla Model S, Tesla Roadster, Nissan LEAF, Chevy VOLT, Mitsubishi iMiev, and Toyota Plug-in Prius are the kinds of highway-ready passenger vehicle models available in Hawaii
3. DBEDT. Monthly energy trends. http://dbedt.hawaii.gov/economic/energy-trends-2/
4. Hawaii Clean Energy Initiative (HCEI). Renewable energy in Hawaii. http://www.hawaiicleanenergyinitiative.org/renewable-energy/
5. Hawaii Energy Policy Forum (HEPF). Hawaii clean energy status report. http://www.hawaiienergypolicy.hawaii.edu/programs-initiatives/clean-energy-metrics/online/3-waiting-challenge/1-transportation-sector.html
6. Maui County Data Book (2012) http://hisbdc.org/Portals/0/MCDB/2012/2012_WholeBook_v2.pdf
7. Hawaii Tourism Authority's annual report. http://www.hawaiitourismauthority.org/default/assets/File/HTA%20AnnuRepFINAL%20WebPosting.pdf
8. Hawaii Tourism Authority. Vision 2014 tourism workforce development strategic plan. The journey to excellence. http://www.hawaiitourismauthority.org/default/assets/File/brand/Fnl-05-08-HTA-16-page-Insert-HB-1.pdf
9. Hawaii State EV. Laws and Incentives. http://energy.hawaii.gov/testbeds-initiatives/ev-ready-program/laws-incentives
10. UH Maui College (2013) Planning for the development of electric vehicle Infrastructure in Maui County. EVs in Paradise, p 29. https://maui.hawaii.edu/eva/home/. Accessed Feb 2013
11. DBEDT (2012) Report to the Maui electric vehicle alliance driving EVs forward: a case study of the market introduction and deployment of the EV in Hawaii. http://energy.hawaii.gov/wp-content/uploads/2011/09/EVReportMauiElectricVehicleAlliance2012.pdf. Accessed Dec 2012
12. Plug-in America (2012) HawaiiI EV ready guidebook for commercial electric vehicle charging station installations. http://energy.hawaii.gov/wp-content/uploads/2011/09/updated-EV-Guidebook_FINAL_Sep-25_2012.pdf. Accessed May 2012

13. UH Maui College (2011) Multi-year plan to integrate high EV penetration with renewable energy on an Island grid system. Proposal to Department of Energy, Clean Cities Community Readiness and Planning for Plug-in Electric Vehicles and Charging Infrastructure. Funding Opportunity Number DE-FOA-0000451
14. High Tech Maui (2014) Maui County continues to lead the way in renewable energy. http://hightechmaui.com/maui-county-continues-to-lead-the-way-in-renewable-energy/. Accessed 28 Jan 2014
15. Hawaiian Electric Company (2012) Clean energy. http://www.heco.com/vcmcontent/StaticFiles/pdf/HawaiiCleanEnergyUpdate92012.pdf. Accessed Sept 2012
16. Altogether four reports were published via Maui EVA. https://maui.hawaii.edu/eva/home/
17. Mead D, Green Tech Media (2012) Top five electric vehicle Initiatives of the year. http://www.greentechmedia.com/articles/read/top-five-ev-initiatives-of-the-year/. Accessed 8 Dec 2011
18. World EV Cities and Ecosystems. http://www.worldevcities.org/
19. Inside EVs (2013) Top 10 and bottom 10 for EV sales in 2013. http://insideevs.com/top-10-and-bottom-10-us-states-for-ev-market-share-in-2013/
20. Navigant Research. Hawaii becoming an EV Paradise. http://www.navigantresearch.com/blog/hawaii-becoming-an-ev-paradise
21. Hawaiian Electric Company (2011) Smart grid demonstration project on Maui. Presented at the Maui EVA kick off meeting
22. JUMPSmartMaui Project.http://www.jumpsmartmaui.com
23. Hitachi (2012) Electric vehicle rental application OSGi and Java in an electric vehicle charging station Infrastructure.http://www.hitel.com/pdf/transport/Electric_Vehicle_Charging_Station_Infrastructure_041312_v4.pdf
24. Claire W, Andy N (2014) Electric vehicle rental services: project in Okinawa, Japan. http://www.cambridgeservicealliance.org/uploads/downloadfiles/Okinawa%20HQP.pdf. Accessed Jan 2014
25. HECO. EV rates and enrollment. http://www.hawaiianelectric.com/heco/Clean-Energy/Electric-Vehicles/EV-Rates-and-Enrollment/Landing/EV-Rates-and-Enrollment?cpsextcurrchannel=1
26. DBEDT (2013) Hawaiian electric companies offer new rates for public EV charging. http://energy.hawaii.gov/wp-content/uploads/2011/09/HECO_EVRates_7.10.13.pdf
27. California Center for Sustainable Energy (CCSE) (2012) Plug-in electric vehicle owner survey. http://energycenter.org/sites/default/files/docs/nav/policy/research-and-reports/California%20Plug-in%20Electric%20Vehicle%20Owner%20Survey%20Report-July%202012.pdf
28. SB 2747 SD1 HD2. Relating to EV parking. http://ssl.csg.org/dockets/2014cycle/34Abills/1434a02hielectricvehicleparking.pdf
29. Honolulu Clean Cities (2012) Lessons learned: the early adoption of electric vehicle charging stations from the perspective of Oahu's commercial properties. http://honolulucleancities.files.wordpress.com/2012/01/lessons-learned-report-for-maui_final_10-22-12.pdf. Accessed Oct 2012
30. UC Berkeley School of Law's Center for Law, Energy and the Environment (CLEE) and UH Maui College (2013) Electric vehicle Paradise: how Hawaii can lead the world in deployment. https://maui.hawaii.edu/eva/home/. Accessed Sept 2013
31. Ku A (2014) University of Hawaii Maui College Final Management Report to Department of Energy. https://maui.hawaii.edu/eva/home/

Charging up Chile: Enabling Shared, Electric Mobility in an Emerging Market

Praveen Subramani

Abstract Santiago de Chile has one of the most extensive and functional metro and bus networks in South America, yet the city is laden with extreme urban congestion and pollution. In this emerging market, where the private vehicle ownership rate is increasing at nearly 7 % annually, electric mobility and vehicle sharing have the potential to significantly mitigate the severe pollution and congestion. However, the high cost of electric vehicle (EV) ownership is far out of reach for the typical Chilean family, whose average net-adjusted disposable income is less than half of the OECD average. This paper proposes an EV sharing ecosystem that creates the opportunity to distribute the high capital cost of EVs across multiple users. In a nation with limited policy incentives for electric mobility, vehicle sharing and strategic partnerships in the private sector involving mining companies, energy providers, automotive OEMs, and research institutions can enable the broader adoption of hybrid and battery electric vehicles. Strategies and recommendations for enabling electric mobility in this emerging but economically divided context are proposed.

Keywords Electric mobility · Electric vehicles · Santiago · Chile · Vehicle sharing · Strategic partnerships · Charging infrastructure · Rapid charging · Latin America · Lithium · Batteries · Mining · Emerging markets

P. Subramani (✉)
Design Lab, Universidad Adolfo Ibáñez, Peñalolén, Santiago, Chile
e-mail: praveen.subramani@uai.cl; praveens@mit.edu

P. Subramani
MIT Media Lab, Massachusetts Institute of Technology, Cambridge, MA, USA

D. Beeton and G. Meyer (eds.), *Electric Vehicle Business Models*,
Lecture Notes in Mobility, DOI 10.1007/978-3-319-12244-1_15

1 Introduction: Chilean Context, Demographics, Existing EVs, and Charging Infrastructure

Chile is a nation of approximately 17 million people, located in the westernmost portion of South America (Fig. 1). With an urbanization rate of 89 % in 2012—higher than that of the USA, UK, France, and the OECD member nation average of 80 %—the cities of Chile face increasing congestion, pollution, and bottlenecks to mobility as their urban populations skyrocket [1]. Its capital city, Santiago de Chile, is home to over 6 million people, accounting for over 35 % of the nation's total population. While Santiago has one of the most extensive and functional public transportation networks in South America, the city is laden with severe urban congestion and pollution. The poor air quality is compounded by the capital's location in a valley, contributing to the accumulation of heavy smog due to an inversion layer. A 2005 study commissioned by Chile's Ministry of the Environment (*Ministerio del Medio Ambiente*—MMA) concluded that over 40 % of the PM_{10} particulate matter in the Santiago Metropolitan Region was attributable to transportation including private vehicles, buses, trucks, and two-wheelers[1] [2]. While the air pollution has lessened somewhat over the past decade due to policy initiatives and increased emissions standards, it still causes severe health and environmental problems for residents of the city as of 2013. For example, during the months of June and July 2013 alone, the regional government issued no less than six preventative environmental warnings, urging residents to stay indoors and reduce driving due to the extremely poor air quality [3].

Multiple governmental agencies in Chile have recognized the potential of electric vehicles (EVs) to contribute to air pollution mitigation and reduction of net energy consumption. To achieve these goals, a target of 70,000 EVs by 2020 was established by a Nationally Appropriate Mitigation Plan (NAMA) and e-Mobility Readiness Plan commissioned by the MMA and the Ministry of Transport and Telecommunication (*Ministerio de Transportes y Telecomunicaciones*—MTT) in 2012 [4]. In addition to the environmental challenges facing Santiago, a number of important economic and technological factors combine to make Chile a particularly compelling candidate for the expansion of hybrid and battery electric vehicles—(H/B)EVs—into the Latin American market. These include the presence of existing EV charging infrastructure (including Latin America's first EV charging station), Chile's history of shared vehicle use in the form of shared taxis and a growing culture around shared mobility, the nation's status as world's largest producer of lithium, a stable and business-friendly government, and the rapidly increasing purchasing power of Chile's citizens.

[1] PM_{10} refers to particulate matter on the order of 10 μ or less that is susceptible to penetrate the deepest parts of the human respiratory system and is a common size delimiter for the measurement of air pollution.

Fig. 1 Location of Chile's capital, Santiago de Chile, within the country and continental South America. Chile is located in the westernmost portion of South America and is geographically isolated by the Pacific Ocean to the west and south, the Atacama Desert to the north, and the Andes Mountains to the east

This paper presents these factors in detail and analyzes each of them in the context of their capabilities to promote electric mobility in the Chilean market, as well as associated strategies for other emerging Latin American nations.

1.1 Jurisdictional Structure

Santiago is Chile's largest conurbation and capital city, though the metropolitan area is actually a collection of 37 separate municipalities, or *comunas*. Figure 2 shows the spatial layout of these municipalities, each of which has its own mayor, town hall, and city administration department. In this paper, references to Santiago indicate the Santiago Metropolitan Region, rather than the specific municipality of Santiago, which lies at the heart of the greater Metropolitan Region. Chile's jurisdictional structure has important implications for electric mobility and transportation initiatives, as the decision-making process is distributed across multiple entities ranging from the local municipal leadership, to provincial/regional leaders, to federal ministries such as MMA and MTT. Three of Santiago's municipalities, which are among the wealthiest in the nation, have played particularly active roles in promoting new forms of sustainable transportation including (H/B)EVs: *Providencia*, *Las Condes*, and *Vitacura*.

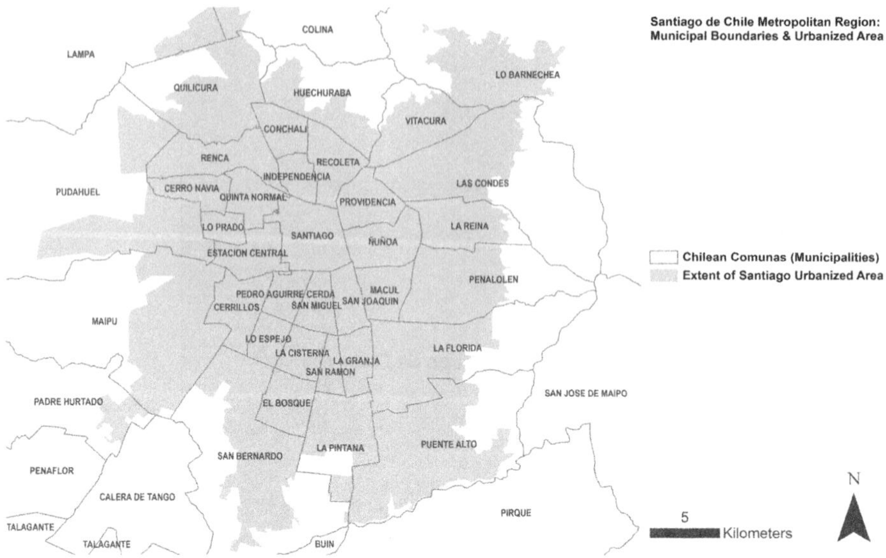

Fig. 2 Map of the Santiago de Chile Metropolitan Region with municipal boundaries, known locally as *comunas*, and *grey shading* indicating the extent of the urbanized area

1.2 Chilean Vehicle Ownership Trends

Since its return to democracy in 1990, Chile has been lauded as a stunning example of rapid economic development and political stability in Latin America. While Chile is only the sixth largest nation in South America by population, it boasts the continent's highest nominal GDP per capita and as of 2013 is the only South American member nation of the Organization for Economic Cooperation and Development (OECD). From 2002 to 2012, Chile's nominal GDP per capita grew over 240 %, from \$4,487 to \$15,363 [5]. Much of the nation's recent economic prosperity can be attributed to its natural resource exports, particularly mining and agriculture. However, Chile is unequivocally still a country in the process of development; its nominal GDP per capita still remained under half of the OCED average of \$36,722 as of 2012. Furthermore, socioeconomic groups in Chile are highly divided and the nation's income disparity, as measured by the Gini coefficient, is the highest among OECD nations. Thus the fruits of Chile's rapidly growing economy disproportionately benefit a limited percentage of Chilean citizens, while much of the country is developing economically at a much slower pace.

The increasing economic affluence of the nation has contributed to the dramatic growth of private vehicle ownership throughout Chile in recent years. The number of passenger cars rose steadily from 88 passenger cars per 1,000 people in 2003 to 127 passenger cars per 1,000 people in 2010, an increase of 44 % in only 7 years (Fig. 3). Due to the high concentration of Chile's population in urban areas, the majority of this rapid growth in passenger vehicle ownership has been concentrated

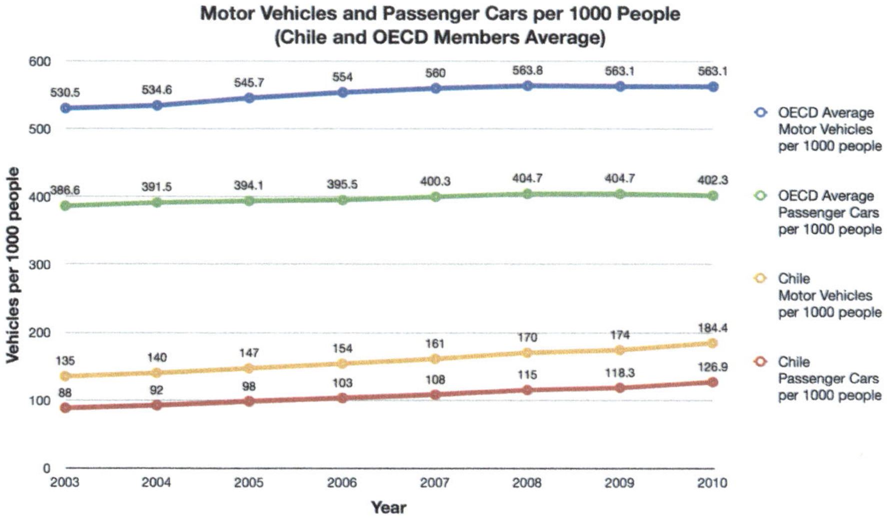

Fig. 3 Motor vehicles and passenger cars per 1,000 people for 2003–2010. Passenger cars refer to road motor vehicles, other than two-wheelers, intended for the carriage of passengers and designed to seat no more than nine people (including driver). Motor vehicles include cars, buses, trucks, and freight vehicles but not two-wheelers [6]

in the cities of Chile, particularly the burgeoning capital of Santiago [6]. Private automobile ownership in Chile is highly aspirational and ownership of a vehicle is considered a symbol of independence, autonomy, and social status. Due to the social importance of vehicles and the increasing purchasing power of Chilean citizens, these trends are expected to continue for the foreseeable future, given 'business-as-usual' policy measures, current gasoline and electricity prices, and transportation infrastructure investment.

1.3 Energy Production Portfolio and Pricing in Chile

To provide context on the logistical and economic factors associated with energy in Chile, a brief summary of energy generation and consumption in Chile is now presented. As of 2013, Chile has over 18 GW of gross installed energy production capacity. Approximately one third of this energy is produced from hydroelectric facilities, an additional third from natural gas, and the remaining third from oil, coal, and renewable sources such as wind and biomass. Chile imports nearly 98 % of the fossil fuels it consumes, so energy security is a recurring subject of national debate and importance. Due to the long and narrow geographic form of the country, which contributes to expansive distances between major urban areas, the nation operates on four separate electricity transmission networks. The *Sistema Interconectado Central* (SIC) grid is the most expansive of the four, providing over 75 %

of the electrical capacity and servicing over 90 % of the population. Energy-intensive copper, lithium, and sulfate mining operations in the northern regions of Chile are serviced by the *Sistema Interconectado del Norte Grande* (SING), which accounts for approximately 23 % of the installed capacity [7].

The cost of electricity in Chile is relatively high, with residential tariffs around USD $0.20–0.24/kWh (depending on the region), due in part to the high reliance on foreign fossil fuel sources such as natural gas, oil, and coal [8]. Gasoline prices in Chile are considerably more expensive than US prices but akin to those of many Western European nations, typically ranging between USD $1.50–2.00/L [9]. The price of residential electricity and gasoline are two important factors to consider when analyzing the feasibility of electric vehicles in any market. Chile's relatively high electricity prices could provide a potential obstacle to the adoption of electric vehicles, so intelligent regulation of EV charging that allows for discounted tariffs in off-peak hours will be an important theme of exploration for energy providers.

1.4 Existing Electric Vehicles, Charging Stations, and E-Mobility Policy in Chile

Approximately 50 battery electric vehicles currently exist in the Santiago metropolitan region, as of late 2013. This figure includes a fleet of 20 Piaggio Porters owned and operated by the municipality of *Providencia*, 16 Mitsubishi i-MiEVs, and a small number of other models including Nissan's LEAF, and BYD's e6. The primary deterrent to wider prevalence of battery electric vehicles has been their high capital cost compared to traditional vehicles. Furthermore, many electric vehicles are sold in Latin America at much higher sticker prices than MSRP for the equivalent or identical product in the USA, Western Europe, or Asia. For example, in May 2011, Mitsubishi launched sales of its i-MiEV battery electric vehicle in Chile at a price of 30 million CLP (approximately USD $59,000). At over double the cost of the sales price for the same vehicle in the USA (USD $29,125), the high price-tag resulted in sluggish sales of only 10 units in 2 years [10]. However, the high sale price of automobiles in Chile is not unique to electric vehicles. Many automobiles of all types sell for 1.2–2 times their retail price in the USA.

Many of Chile's OECD peers including the US and many EU nations offer some degree of governmentally supported incentives (often in the form of tax rebates, reduced road tolls, or reduced vehicle registration fees) for early adopters of (H/B) EVs. For example, the UK's Plug-in Car Grant offers a 25 % grant (up to GBP £5,000) towards the cost of new plug-in vehicles for both consumers and fleet operators. Similarly, residents in the state of California, USA can currently take advantage of up to $7,500 in federal tax credits and up to $2,500 of state-sponsored rebates for purchasing certain battery electric vehicles [11]. The only similar policy measure in Chile even reminiscent of these proactive subsides by other OECD nations was a 2008 hybrid vehicle law which waived annual vehicle registration fees for a period of 4 years for purchasers of hybrid vehicles. However, this is a relatively small fee compared to the high prices of hybrid and battery electric vehicles in Chile, so the law was credited only with contributing to the registration of approximately 400 hybrid vehicles from 2008 to 2010 [4].

Despite the lack of large-scale policy incentives such as purchase rebates, the business-friendly Chilean government has provided some small contributions to the promotion of electric mobility in the nation. For example, the government has sponsored innovation seed grants to new businesses committed to electric mobility that seek to enter the market. Growing concerns over air pollution in the capital and recognition of the impact of Chile's lithium abundance have driven local entrepreneurs to contribute to promotion of an e-mobility ecosystem. For example, a Chilean vehicle startup Voze has developed a fully electric three-wheeler vehicle prototype, named *Lüfke*. The three-wheeler was designed and developed in Chile with an emphasis on urban environments, and the company aims to take their product to market in 2014. The company was financed in part by innovation seed grants from CORFO, a public-sector organization dedicated to promoting entrepreneurship, innovation, and economic growth in Chile [12]. Similarly, CORFO and the MMA have been active supporters of projects such as *Desafio Cero*, an annual competition and road race of zero-emission compact urban vehicles. While these initiatives and others have created a positive impact in terms of public relations and visibility of electric vehicles, they appear to have had little concrete impact in the actual adoption rate of (H/B)EVs in Chile [13].

1.5 Electric Charging Infrastructure in Santiago de Chile

A crucial factor for promoting the adoption of electric mobility in Chile is the installation of public infrastructure for vehicle charging. This process is well underway in Santiago and has been driven largely by private sector investment. In 2011, Chile became the first nation in Latin America to install a publicly accessible electric vehicle charging station. Most stations are owned and operated by *Chilectra*, the private-sector electricity distribution utility that has continued to play a pivotal role in the installation of charging infrastructure throughout the city. As of October 2013, four publicly accessible standard charging stations (AC 3.5 kW, based on the SAEJ1772 standard) and four rapid charging stations (DC 50 kW, based on the CHAdeMO standard) have been installed in the Santiago metropolitan area. Some of these stations even provide charge plugs for multiple vehicles. *Chilectra* reports that the charging stations are largely underutilized, given the current absence of a significant number of plug-in EVs in the Santiago metropolitan area [14].

Some of these charging stations offer free charging while others charge a nominal electricity fee to users of 100 CLP (approximately USD \$0.20)/(kWh) of charge, similar to the residential tariff. Thus a nominal 16 kWh charge for a vehicle such as the Mitsubishi i-MiEV would cost 1,600 CLP (USD \$3.25). Furthermore, a 30 % discount is offered during off-peak charging times such as night hours to incentivize the use of chargers [15]. Rapid charging is billed by time, with a fixed price of 2,000 CLP (USD $4.00)/15-min charging interval [16].

Infrastructural connectivity and the availability of electric transformational capacity has not thus far been a bottleneck for electric mobility in Chile. According to the February 2012 NAMA, *Chilectra* has stated that with current generating capacity, approximately 200,000 EVs (corresponding to a consumption of over 400 GWh/year) could be charged without necessitating the installation of additional capacity. Furthermore, the widespread availability of 220 V single-phase power in Chilean residences provides logical connection points for home charging stations in the Level I and II ranges, which can be billed at standard residential tariffs with discounts for off-peak and overnight charging [4].

The placement of public electric vehicle charging infrastructure in Santiago to date has been largely motivated by proximity to high-income neighbourhoods where EV owners are more likely to reside, and have been placed at points-of-service such as malls and gas stations with high visibility. For example, the distribution of existing EV charging infrastructure has been heavily concentrated in the high-income municipalities of *Vitacura* and *Las Condes*, with five of the eight charging stations located in these areas, which together are home to only about 5 % of Santiago's population. The existing public charging stations are also mostly located in areas with relatively low population density, with the exception of one centrally located station in a high-density area, as indicated in Fig. 4. According to *Chilectra* representatives, station placement is determined by site availability and

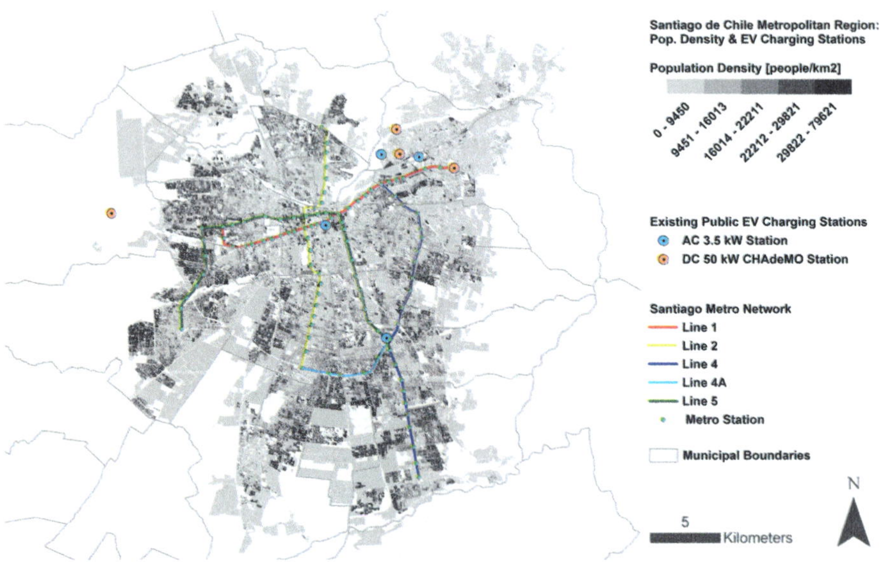

Fig. 4 Santiago Metropolitan Region population density (shown in people per square kilometre and classified by quantile), the metro network, and existing electric vehicle charging stations. Many of the existing public charging stations are clustered in the high-income areas of the city in the northeast that are home to more EV owners, but exhibit relatively low population density

visibility of the installation, proximity to residences of existing EV owners, and transit/activity corridors such as highway rest stations and malls. Section 3 presents an accessibility study of existing EV charging stations and data-driven recommendations for placement of future infrastructure.

2 Vehicle Sharing as a Cost Reduction Measure for Enabling Electric Mobility

In addition to the lack of extensive public charging infrastructure on a regional scale, the high capital cost of electric vehicles has proven to be a significant deterrent to the widespread adoption of EVs worldwide. This is particularly true in the Chilean market, as the average household net-adjusted disposable income is $11,039/year, less than 50 % of the OECD average of $23,047 [17]. Consequently, automobile ownership is highly aspirational for many low and middle-income Chilean families and EVs (which often carry sticker prices of well over $30,000) are generally only financially accessible to high-income households.

One of the most promising strategies for the broader introduction of (H/B)EVs into the Chilean market is the use of vehicle sharing to spread the cost of a single vehicle across multiple users and increase consumer exposure to and comfort with EV technology. Two distinct forms of vehicle sharing are presented in this analysis: (1) implementation of (H/B)EVs in fleets of *colectivos*, or shared taxis, and (2) implementation of (H/B)EVs in fleets of shared vehicles that are available for hourly or daily rental by the general public.

2.1 Colectivo Scenario (Dedicated Driver, Payment Per Trip)

Although private vehicle ownership is highly aspirational in Chile and other developing nations, Chile and many of its South American neighbours have a strong history of centralized vehicle sharing in the form of the *colectivo*, or collective taxi. A *colectivo* is a type of shared taxi service with relatively fixed routes that typically depart when all seats of the vehicle are filled. *Colectivos* will often deviate slightly from their route to drop passengers to specific locations, within a reasonable range, and allow each rider to pay comparably less than they would pay for the same trip in a single-occupancy taxi. These shared taxis compose an important part of the transportation fleet of cities such as Santiago, where many communities (particularly those on the outskirts of the city) are underserviced by the public transit network consisting of the metro and buses. *Colectivos* are a uniquely Latin American urban transportation solution that also cater to users who require more direct, comfortable, or rapid mobility than provided by public transportation, but do not own (or prefer not to use) a private automobile.

Colectivos represent an important and unique opportunity for the introduction of electric mobility pilot programmes in Chile and other South American nations. Gasoline and vehicle maintenance are the primary expenditures of *colectivo* operators so the prospect of reducing the per-kilometer cost of the routes (albeit with a higher up-front capital cost for an EV) could be extremely appealing for these vehicles that log significant mileage in a typical day. Due to their relatively fixed routes, drivers of *colectivos* would be able manage the range of a BEV more accurately than a driver of a typical taxi or passenger vehicle. Furthermore, as *colectivos* are a common transportation mode for Chileans without private vehicles, the use of EVs in *colectivo* fleets could improve exposure and knowledge of the technology among a demographic that may purchase private vehicles in the future as their socioeconomic status permits. Finally, as *colectivos* typically operate in fleets with queues at specific transportation hubs, drivers often have significant stop-time while waiting for their turn to carry passengers. This waiting time offers an opportunity for recharging of plug-in vehicles, particularly if rapid charging options are available [4].

2.2 Traditional Vehicle Sharing Scenario (User-Driven, Payment by Rental Time or Mileage)

Conventional vehicle sharing programmes, based on the European or North American model employed by companies such as Zipcar, Car2Go, and BeMobility will also assist in the creation of new business opportunities for electric mobility in Chile and its Latin American neighbours. Several existing initiatives involving vehicle sharing are already active in Santiago, though none yet incorporates (H/B) EVs. The increasing visibility of vehicle sharing programmes and initiatives in Chile is gradually assisting in the creation of a socioeconomic and cultural class that has already been extensively documented in Western Europe and North America: young professionals who prefer the flexibility, cost-reduction, and urban mobility advantages offered by shared vehicles to owning a dedicated private vehicle.

Prior to 2013, the municipality of *Providencia* offered the only bike sharing system in Santiago and employs unautomated infrastructure. Unlike most of its North American and European counterparts, *Providencia's* bikesharing programme is not kiosk-based and is largely manually-operated. The bikes do not lock to the station, but are instead monitored by an attendant. This attendant also manages the rental process, using a handheld electronic device. To rent a bicycle, users approach the agent who enters the user's national identification number into the device, and associates it with an ID number that is physically painted on each of the bikes. Upon return, users report their ID number and bike number at the drop-off station to the attendant, who marks it as returned in the system. The agents are present until about 9:00 p.m. at night, after which collection trucks are circulated, the bikes are removed, and locked in a storage facility for the night. Users can register at each of the stations for CLP 2,000 (about USD $4) per month or CLP 15,000 (about USD $30) annually. While the process seems cumbersome and inefficient compared to automated, kiosk-based bikesharing systems, it is a reflection of the fact that human capital and labor are typically less expensive in Chile than physical products and commodities. Thus it may be more cost-effective for the *Providencia* local government to equip each station with an agent as opposed to investing in electronic bikesharing kiosks and automated rental infrastructure, which also require regular maintenance and service [18].

In March 2013, the MMA announced the launch of an automated kiosk-based bikesharing programme (similar to those employed by Sao Paulo and Rio de Janeiro in their bikesharing programmes) funded by Banco Itaú Chile and B-cycle LLC, a Wisconsin-based provider of bikeshare infrastructure. While the programme is nascent, it aims to expand the infrastructure through multiple municipalities of Santiago over 4 years to reach a capacity of 3,000 bicycles and 300 stations, involving a total investment of USD $15 million [19]. The first pilot stations of the kiosks were installed in *Vitacura* in October 2013. As this more comprehensive bike-sharing programme expands and the concept of shared vehicles grows more amenable and visible to residents of Santiago, additional pilots with electric bicycles or scooters will offer an important opportunity for integrating electric mobility with vehicle sharing.

Strategic partnerships with charging infrastructure providers and automotive OEMs seeking to boost their EV fleet presence in the nation will also be critical to ensuring the viability and success of a conventional vehicle sharing programme. Pilot fleets of shared vehicles connecting corporate campuses of electric infrastructure companies such as *Chilectra* represent an important opportunity for nearly immediate installation of fleet EVs, with potential expansion to a larger user base. A key advantage of piloting shared vehicles in commercial fleets is the comparatively high levels of trust and responsibility within a corporate community (compared to the general public) and the relatively limited origin/destination matrix for such fleets. Commercial partners that could target the introduction of vehicle sharing with integrated (H/B)EVs include corporations such as banks, automotive OEMs and distributers, real estate developers, infrastructure service providers, ITC companies, and EPC firms.

In addition, governmental vehicle fleets such as those of Santiago's numerous municipalities also present an important opportunity for the integration of (H/B) EVs. For example, the municipalities of *Providencia* and *Vitacura* currently operate fleets of municipal vehicles that include several (H/B)EVs. These fleets contribute to "environmentally-friendly" branding of the municipalities, reduce air pollution, and increase public awareness of the technology. Due to the jurisdictional structure in Chile (described in Sect. 1.1) in which each municipality is responsible for urban services such as waste collection, parking enforcement, and some transportation infrastructure, neighbouring municipalities tend to be quite competitive as they vie for visibility and real-estate investment. As such, many of the higher-income municipalities have already invested heavily in 'green' technologies and sponsored showcases of sustainable transportation initiatives for their municipal fleets and urban services.

Broadening the penetration of (H/B)EVs into the Chilean and other Latin American markets through vehicle sharing offers many key advantages. Primarily, it allows end-users to test and acclimate to driving (H/B)EVs without assuming the capital cost of the entire vehicle. As electric vehicles are currently neither abundant nor highly visible in Chile, many locals have doubts about the feasibility of EV technologies to provide adequate range, climb hills, and operate reliably. Thus vehicle sharing allows users to familiarize themselves with the technology for a relatively low cost and alleviate their fears of range anxiety, poor performance, or other technological problems associated with EVs. Secondly, vehicle sharing is documented worldwide to appeal heavily to the key demographic of increasingly affluent young professionals in the 18–40 age range, who are often seeking additional mobility options and transportation flexibility. Targeting this demographic can contribute to the development of brand/technology loyalty and acclimatization to different modes of mobility such as shared vehicles. Furthermore, this demographic is typically among the most technologically connected and already has widespread access to the Internet, mobile devices, and other interfaces that allow for easier integration of vehicle sharing with other transportation modes. Finally, vehicle sharing has been indicated to significantly contribute to the reduction of congestion and pollution in urban areas due to its ability to supplant private automobiles as a mode of transportation and reduce urban space allocated for parking. Pairing vehicle sharing with (H/B)EVs is a logical strategy for cities such as Santiago that face daunting challenges of air pollution and resource consumption. In addition to the environmental benefits, critical branding and marketing opportunities around environmental friendly technologies can help attract private sponsorship and governmental investment in such initiatives.

3 EV Charging Station and Metro Station Accessibility Analysis

As previously mentioned, the current site selection and installation process for public EV charging infrastructure in Santiago is conducted principally by *Chilectra* and its partners. Station placement is governed largely by visibility and proximity to residences of existing EV owners, which tend to be in high-income neighbourhoods. This section presents an analysis linking demographic, income, and transportation data with EV charging station site suitability. In addition, recommendations for the integration of vehicle sharing including (H/B)EVs with the public transportation network are presented.

Using demographic and transportation data gathered from data sources including the Chilean census, household transportation surveys, an accessibility analysis was conducted to motivate the placement of future EV charging and vehicle sharing stations in the Santiago Metropolitan Region. The following methodology could also be applied to other Latin American cities, assuming availability of similar datasets. The studies presented in Sects. 3.2 and 3.3 represent travel time impedance analyses, which are defined by the gravity model of trip distribution as described by Levinson and Kumar [20]. These impedance analyses leverage a data set of network travel times, specifiable for pedestrian walking times, public transportation trip times, and private automobile driving times. The data set was generated and provided by the Center for Territorial Intelligence (*Centro de Inteligencia Territorial*—CIT) at the Universidad Adolfo Ibáñez.[2] The resulting impedance analyses indicate spatial accessibility to distinct points (i.e. metro stations or EV charging stations) based on specific modes of transportation such as walking and driving [21].

3.1 Metro Station Location and Household Income

Initially, the spatial correlation between metro station location and household income was analyzed to highlight the importance of proximity to transportation nodes in influencing land value (perceived or real). Of particular interest is the strong positive correlation between proximity to metro lines higher income households, as illustrated in Fig. 5. While the higher-income regions of the city are largely clustered in the northeast regions in municipalities such as *Providencia, Las Condes*, *Vitacura*, and *Lo Barnachea*, an interesting trend of higher-income residences located along the metro corridors is evident. This is particularly observable on the southern portions of Lines 2 (yellow) and 4 (indigo). These lines cut into largely low-income areas, while the land tracts abutting the metro lines continue to attract higher-income residents. Due to the relative recentness of vehicle sharing in Santiago, data on the spatial correlation between vehicle sharing stations and household income is not yet available. However, it is possible that access to vehicle sharing stations ranging from bike sharing stations to carsharing stations may demonstrate a similar correlation, if these modes are adopted by local residents as a regular form of urban transportation.

3.2 Metro Station Pedestrian Accessibility (5, 10, 15 min) and Household Income

Based on the gravity model, travel time impedance analysis was initially conducted for 5, 10 and 15 min walking times originating at Santiago metro stations. The resulting analysis reveals a number of insights relating demographic data to transportation accessibility (Fig. 6). The existing public transportation network is

[2] Travel time impedances were generated from a dataset of combined measurement and simulated information provided by the CIT. Demographic data regarding household income is drawn from the 2002 Chilean census, while population density is from a 2010 dataset. The most recent 2012 Chilean census is widely considered an unusable dataset, as it failed to account for nearly 10 % of the population due to a severe error in which homes were wrongly labeled as empty.

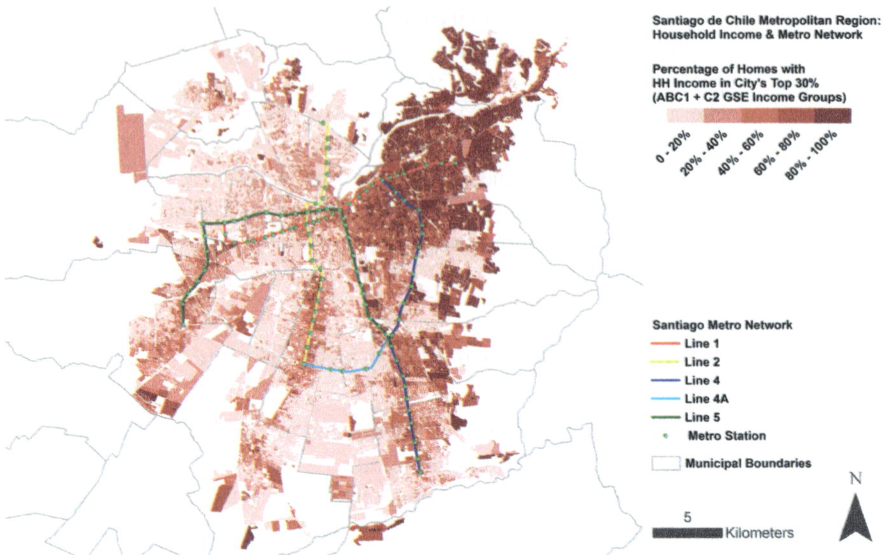

Fig. 5 Percentage of households in *top* two income tiers in Santiago de Chile classified by quantile (Socioeconomic groups ABC1 and C2, which correspond to incomes in the *top* 30 %), and the city's metro network. A strong positive correlation exists between income and areas of land that are in close proximity to metro stations. This is particularly evident on the southern portions of Lines 2 (*yellow*) and 4 (*indigo*) and the western portion of Line 5 (*green*), which cut into largely low-income areas while the land tracts abutting the metro lines attract higher-income residents

anchored by the Metro de Santiago, which includes five lines and two additional currently under construction. While the Santiago metro system is South America's most extensive, the walking accessibility study indicates that the metro does not provide adequate coverage for significant portions of the urbanized area, particularly the high-density areas on the outskirts (see Fig. 8 for superimposition of accessibility over population density).[3] This analysis indicates that only 15 % of Santiago's urbanized area (94 km^2 of the city, out of the total urbanized area of 640 km^2), is within a 15-min walking time from an existing metro station.

Considering vehicle sharing stations as a complementary asset to the public transportation network, there is a clear opportunity for vehicle sharing stations to fill in the extensive gaps in transportation coverage. Installation of vehicle sharing is typically less costly than interventions such as metro networks, which require time and resource-consuming installation of major infrastructure. Vehicle sharing with conventional vehicles could thus contribute to the improvement of transportation accessibility in the Santiago Metropolitan Region. Vehicle sharing, paired with

[3] Santiago's public transportation system also includes an extensive network of local buses, which were not considered in this study.

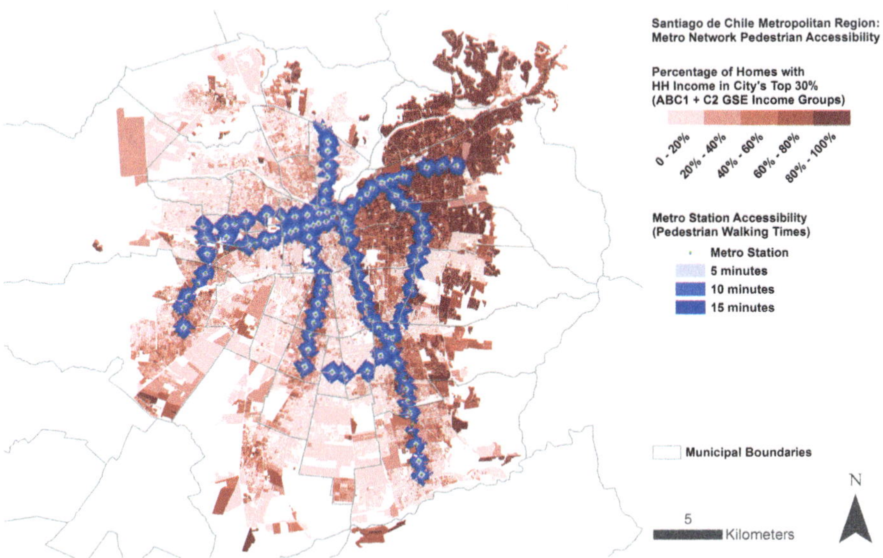

Fig. 6 Pedestrian accessibility of metro stations in the Santiago Metropolitan Region, superimposed on the map of household income distribution. The accessibility analysis was conducted for pedestrian walking times and shows 5, 10, and 15-min walking travel times from each station. This analysis indicates that 85 % of the urbanized area, including both high-income and low-income areas, are not within a 15-min walk of Santiago's metro network

(H/B)EVs, could further contribute to the reduction of the severe air pollution that plagues the city due to the commensurately lower GHG emissions of these vehicles. Given the strong governmental and societal interest in reducing urban air pollution, integration of (H/B)EVs is particularly compelling in the case of Santiago.

3.3 EV Charging Station Driving Accessibility (5, 10, 15 min) and Household Income

A travel time impedance analysis was conducted for private automobile driving times with trips originating at the eight existing electric vehicle charging stations in the Santiago Metropolitan Region (Fig. 7). The resulting analysis indicates that existing accessibility is quite high for high-income regions of the city, particularly in the municipalities of *Vitacura* and *Las Condes*, which are home to five of the eight (62.5 %) public charging stations in the city, but less than 5 % of the population. The westernmost station connects to Chile's Route 68 and provides rapid charging access to regional travellers to neighbouring coastal cities such as *Valparaiso* and *Viña del Mar*. Only one station, housed at the offices of *Chilectra* near central Santiago, is within a 15 min driving distance of the central and densely populated portions of the city. Other parts of the city such as the lower income portions in

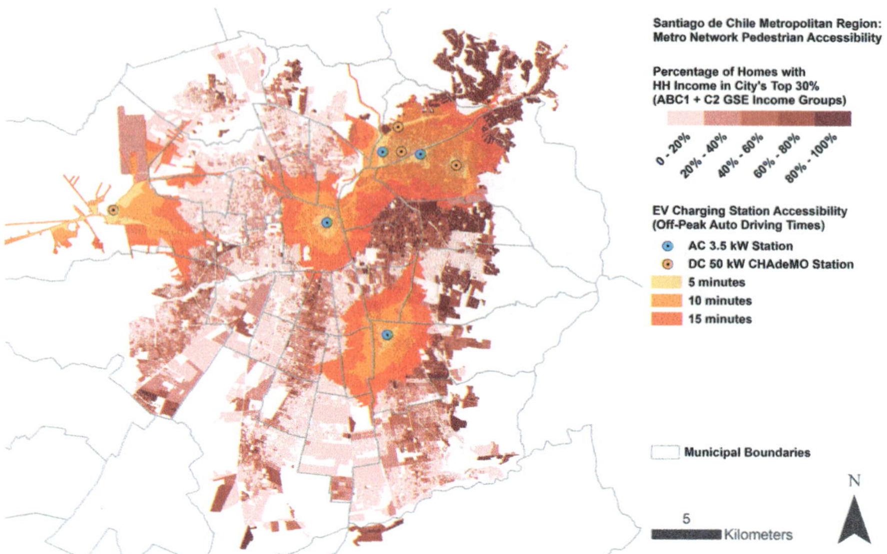

Fig. 7 Accessibility of electric vehicle charging stations in the Santiago Metropolitan Region, superimposed on the map of household income distribution. The accessibility analysis was conducted for automobile driving times in off-peak conditions and shows 5, 10, and 15 min driving areas from each station. The presence of five public charging stations in the northeastern (high-income) portion of the city provides ample charging station access in this region, while other portions of the city exhibit lower accessibility. However, the network demonstrates significant coverage overall: 47 % of Santiago's urbanized area is within a 15-min off-peak driving time from an existing charging station

southern and northwestern Santiago exhibit considerably lower accessibility to all charging stations. These are also regions that suffer from lower general transportation accessibility, which could be augmented with vehicle sharing programmes. In general, driving time accessibility to EV charging stations in Santiago is high among Latin American nations. This analysis indicates that 47 % of Santiago's urbanized area (299 km^2 of the city, out of the total urbanized area of 640 km^2), is within a 15-min off-peak driving time from one of the eight charging stations.

3.4 Holistic Urban Accessibility and Future Charging Station Opportunities Analysis

The expansion of future charging infrastructure, particularly for shared vehicles and *colectivo* fleets, can leverage this data-driven travel time impedance and demographic mapping methodology to deploy charging infrastructure that covers gaps in existing service areas as opposed to duplication of services, as is largely present in the northeastern portion of the city. Superimposing combined metro station and charging station accessibility over a map of population density further highlights

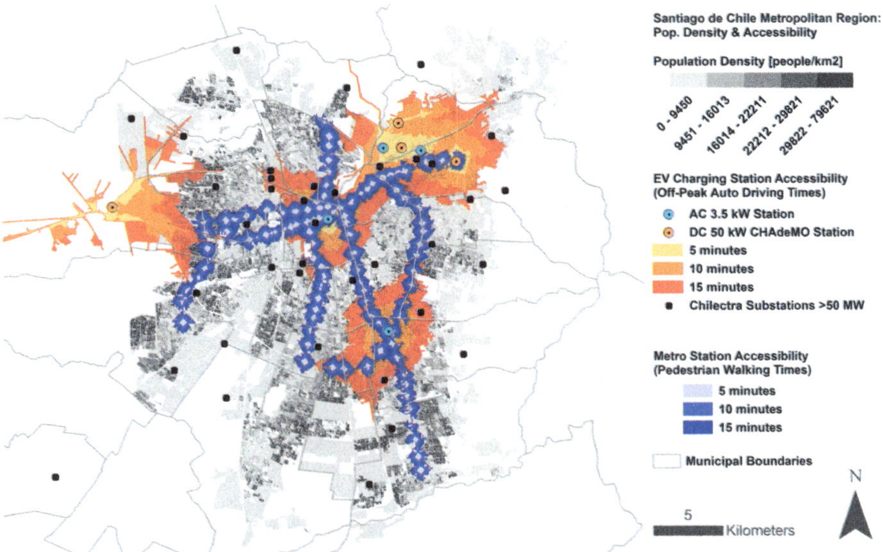

Fig. 8 Combined metro station and EV charging station accessibility in the Santiago Metropolitan Region superimposed upon population density. This holistic map indicates several opportunities for interventions of vehicle sharing stations including (B/H)EVs in densely populated areas of the city. While these networks service some of the densest areas of the city, others (particularly in low-income neighbourhoods on the outskirts) exhibit reduced accessibility. *Chilectra* urban substations with over 50 MW of transformational capacity are also shown, indicating potential points of connectivity for major installations of EV charging infrastructure, such as mobility hubs incorporating significant numbers of shared electric vehicles

the need for improved transportation accessibility in the outskirts of the city, where many high-density communities are largely underserved by the metro but could potentially connect to *colectivo* routes, bus routes, or shared vehicle systems that include (B/H)EVs. The spatial mapping of major electric transformational infrastructure, such as substations with capacity of over 50 MW also indicates opportunities for inclusion of additional EV charging infrastructure which can form the basis for multi-modal vehicle sharing stations that include traditional and electric vehicles (Fig. 8).

4 Lithium Production and Electric Mobility

The final factor presented in this analysis of opportunities for integration of electric mobility is the role of Chile's lithium mining companies and its connection to the burgeoning EV battery industry. Chile is the world's largest producer of lithium, an essential component of EV batteries, accounting for over 35 % of global lithium

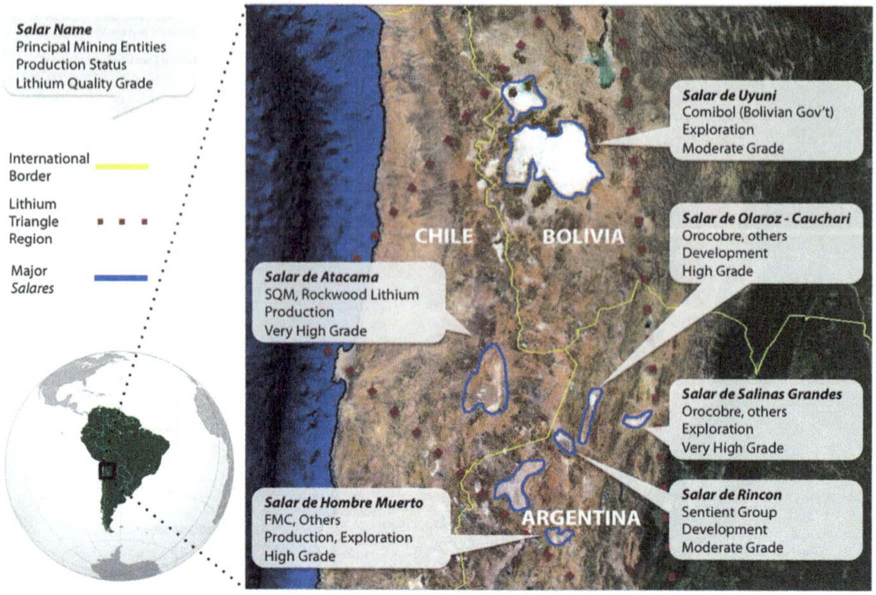

Fig. 9 The 'Lithium Triangle', an incredibly lithium-rich region of South America that is responsible for the largest portion of global lithium extraction. A significant portion of the Lithium Triangle lies within Chilean territory and Chilean lithium sourced from the *Salar de Atacama* is considered among the highest grade of global lithium reserves [23]

mining production in 2012, excluding US production[4] [22]. In a nation with practically nonexistent policy incentives for electric mobility, but whose economy and politics are heavily influenced by the mining industry, investment in end-use applications for lithium-dependent technologies such as battery electric vehicles by mining interests offers an additional important opportunity for the promotion of electric mobility.

Two private entities currently dominate the extraction and processing of lithium in Chile: *Sociedad Quimica y Minera* (SQM) and Rockwood Lithium—which produce lithium carbonate, lithium hydroxide, and lithium chloride. Of these compounds, lithium carbonate is of principal interest to the automotive battery industry, as it is used in the construction of lithium-ion battery cathodes. Lithium is extracted principally in a region known as the Lithium Triangle, which includes territories of Chile, Argentina, and Bolivia (Fig. 9). Currently the majority of this lithium is exported to East Asia, the USA, and the EU for use in advanced manufacturing and research and development (R&D). While Chile has a well-established mining sector, R&D activities around lithium-based technologies are still quite nascent, so lithium is treated only as a commodity item [23].

[4] The US Geological Survey Mineral Commodities Summary for 2013 excludes data about US mineral production by law to avoid disclosing company proprietary data.

A brief summary of the lithium mining process and the steps involved in processing the material for battery applications is presented and illustrated in Fig. 10. The fabrication of lithium-based batteries for EVs requires technical-grade lithium carbonate (Li_2CO_3). In the salt flats or *salares* of the Lithium Triangle, brine is pumped from below the salt crust and deposited in shallow pools with large surface areas. Through a gradual process of solar evaporation, various salts begin to precipitate from the brine. The raw extracted salt brine in Chile's *Salar de Atacama*, one of the world's largest reserves of high-grade lithium, has an initial lithium content of 0.2 % and reaches nearly 6 % after the solar evaporation process. However, the resulting solution contains significant impurities of magnesium, boron, and sulphate. Consequently, the concentrated lithium brine is transported by tanker truck to processing plants in the regional capital of Antofogasta and undergoes a purification process by precipitation of lithium carbonate. The post-processing purification is close to 99.5 %, above the specified minimum battery-grade standard of 99.1 %

Numerical estimates of the quantity of lithium carbonate required per kWh of energy capacity for an automotive-grade battery varies heavily amongst industry estimates and published literature. A highly cited 2012 USGS analysis of lithium use in batteries estimates that between 117 and 250 g of lithium equivalent are required per kWh, depending on a variety of factors such as the anode type and the specific battery chemistry characteristics. A single gram of lithium corresponds to 5.32 g of lithium carbonate, as governed the stoichiometric ratio of the compound. Thus, between 622 and 1,330 g of lithium carbonate are required per kWh of automotive grade battery. Using a 25 kWh automotive battery as a reference size, this amounts to between 15.5 and 33.3 kg of lithium carbonate for an electric vehicle, making electric vehicle batteries one of the most significant markets for technical-grade lithium compounds in the near future [24].

Chilean mining industry leaders have long considered lithium and its related compounds as commodity products, which are exported at relatively low costs to commercial entities in foreign nations that engage in advanced technology development with lithium. Lithium applications range from batteries to ceramics to

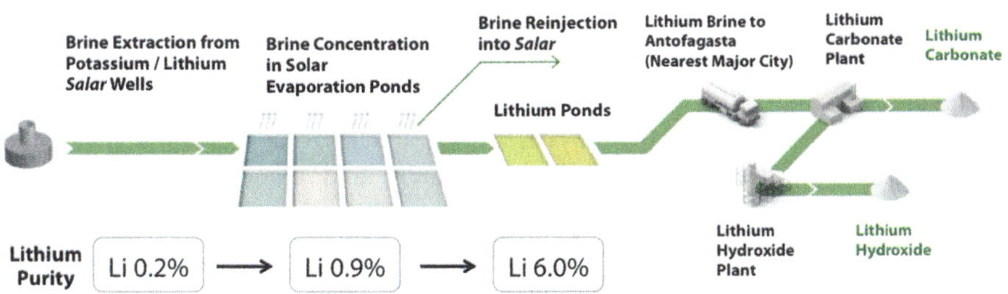

Fig. 10 Lithium extraction and purification process in Northern Chile's Atacama Desert. Electric vehicle lithium-ion batteries require technical grade lithium carbonate of over 99 % purity. Image modified from original with permission from Prof. Miguel Herrera, UAI Mining Department and SQM

industrial lubricants. Until very recently, little advanced technology based on lithium has been developed in Chile or anywhere in Latin America. Thus, the automotive-grade battery-manufacturing ecosystem will continue to be dominated by eastern Asian nations, the USA, and the EU for the foreseeable future.

As of 2013, only one notable research initiative exists around the topic of exploring advanced research applications for lithium-based technologies in Chile. The Lithium Innovation Center (*Centro de Innovación de Litio*—CIL), a strategic initiative consisting of SQM, Rockwood Lithium, Japan's Marubeni (the Chilean distributer of Nissan vehicles in Chile), and the University of Chile that supports research and development activities involving end-use applications of lithium such as electric vehicles and the development of batteries. In July 2013, the CIL announced the creation of the first Chilean-made lithium battery suitable for electric mobility, based on a LiNiMnCo chemistry. The project was funded fully by private resources with no state support, indicating the increasing importance of advanced lithium based technologies to the consortium members from the private sector.

The creation of this consortium in the context of Chile's mineral-wealth is significant for a number of reasons. Primarily, it is an indication of private sector entities assuming increasingly active roles to contribute to the adoption of electric mobility and energy storage without governmental support. Secondly, it is a rare example of an academic-industrial R&D collaboration in Chile, a nation that has largely been economically driven by natural resource and commodity exports while investing relatively little in advanced research. Finally, the consortium has actively contributed to bringing EVs and electric mobility into the public spotlight with a number of high profile projects, potentially reducing fear of new technologies and increasing comfort levels among the general public. Nonetheless, significant shifts in mindset among leadership of mineral extraction companies in Chile will need to occur to arrive at a point in which these companies truly promote electric mobility in a more comprehensive manner.

Potential long-term business strategies include investment in advanced research and development of batteries and energy storage technologies for EVs and local subsidies or discounts on lithium-rich products. Marketing and branding business campaigns can also play a role in the development of Chile's lithium industry and its ties to electric mobility. Perhaps one of the most notable campaigns for promoting commodity-product marketing towards consumers was Intel Corporation's "Intel Inside" advertising campaign, which promoted the semiconductor and integrated circuit manufacturer's technology directly to end-users who were purchasing a more complex product such as a personal computer or laptop. Similar strategies can be employed by EV battery manufacturers and even lithium companies, allowing them to market their product and branding directly to consumers and potentially increase demand for the base commodity product. Regional-specific campaigns, which are already predominant in the agricultural industry in Chile (i.e. agricultural products that are clearly and proudly marketed as products of Chile), are also possible, such as explicit branding of Chilean-source lithium, which is already regarded within the industry as among the world's highest quality.

5 Conclusions and Extension to Other Latin American Markets

The introduction of electric mobility into emerging markets, including those of Latin America, will be a gradual process that will require support from a number of entities. Unlike in the USA, EU, and China, however, this support has and may continue to come more from the private sector than in the form of government incentives for (H/B)EVs. In particular, the high capital cost of electric vehicles is the most significant deterrent to their broader adoption in both emerging and developed markets. In the case of Chile, lack of government incentives for low-emission vehicles has created an environment in which private industry has taken the leadership in enabling electric mobility on a limited scale. This paper presented and evaluated a number of factors that will likely play critical roles in the penetration of electric mobility into the Chilean market. These factors include a rapidly growing economy and increased purchasing power of citizens, private industry deployment of charging infrastructure, the potential of vehicle sharing to enable access to and reduce public apprehension of electric vehicles, and the importance of Chile's lithium mining companies seeking closer alignment with the automotive battery manufacturing industry.

While the adoption of electric vehicles in Chile is still in a nascent stage, it is evident that the nation is emerging as a leader in Latin America regarding the installation of charging infrastructure and investment in electric mobility. For example, the analysis presented in Sect. 3 concluded that 47 % of the land area of Santiago already lies within a 15-min off-peak driving time from an existing electric vehicle charging station. Most importantly, Chile could eventually function as a global incubator for electric mobility initiatives that operate without government subsidy. Private entities such as *Chilectra* and local automotive OEMs, which have a vested interest in promoting electric vehicles as a wider-adopted mode of transportation, have played a critical role in investment and innovation in the space of electric mobility. This private-sector investment has been essential, and has served as the primary catalyst for electric mobility in an incentive-averse political context.

The vehicle sharing case study presented in Sect. 2 offers one of the most promising short-term solutions for introducing more electric vehicles to the Chilean market. Vehicle sharing offers the important advantage of distributing the high capital cost of EVs across multiple users and allows consumers to grow comfortable with the technology without assuming the cost of a private vehicle. *Colectivo* fleets provide a uniquely Latin American opportunity for integration of (H/B)EVs into urban transportation fleets that operate along relatively fixed routes and cover gaps in the existing public transportation network. Furthermore, data-driven placement of EV charging infrastructure and vehicle sharing hubs can be conducted as demonstrated in Sect. 3 to improve transportation accessibility, reduce air pollution, and provide additional sustainable mobility options that may encourage some users to avoid purchasing a private automobile for their daily transportation needs, or prevent the purchase of a second vehicle among higher-income families.

The Chilean example provides a particularly interesting model for other emerging markets in Latin America, in which governmental incentives for electric mobility are often weak or non-existent. Following Chile's example, other private entities Latin American nations can follow the charge of expanding the markets for (H/B)EVs through private sector investment. Strategic partnerships between energy providers, automotive OEMs, and ICT companies will be critical to drive electric mobility forward in an effective manner. Vehicle sharing and other forms of fleet vehicles such as taxi and corporate fleets also present important opportunities for the deeper integration of electric vehicles throughout Latin America. Other Latin America nations, such as Colombia, have already begun small pilots of electric taxis to improve visibility and comfort with EV technology. In Chile and throughout the region, strategic initiatives from the private sector and vehicle sharing including *colectivos*, taxi services, and fleet vehicles are proving to be the most important mechanisms for the introduction of (H/B)EV technologies into the diverse mixture of urban transportation modes in Latin America.

Acknowledgments This research was funded in part by the Fulbright Commission of Chile (www.fulbrightchile.cl), to which the author expresses much gratitude. The author also expresses sincere thanks to colleagues at the Design Lab and Center for Territorial Intelligence of the *Universidad Adolfo Ibáñez* (UAI) including Mr. Ricardo Truffello, Dr. Luis Valenzuela, Dr. Sergio Araya, and Dr. Alexandros Tsamis for access to geospatial demographic data and guidance on the research project in the local context. Dr. Miguel Herrera of the UAI Mining Department provided important information and contacts regarding the process of lithium extraction and processing in Chile. Ms. Andrea Terrazas and Ms. Carolina Busquets of the UAI Design Lab provided invaluable administrative and logistical support of the project. Dr. Boyd Cohen of the *Universidad de Desarrollo* contributed significant insight on the feasibility of vehicle sharing in Chile. Dr. Jaime Alee of the CIL provided insight into applications of Chilean-mined lithium to the transportation industry. Finally, relevant data on existing EV presence, location of existing charging stations, and major electrical transformational infrastructure were provided by Mr. Orlando Meneses and Mr. Cesar Sanchez of *Chilectra*'s Innovation Division.

References

1. The World Bank (2012) Urban population (% of total). United Nations, World Urbanization Prospects. http://data.worldbank.org/indicator/SP.URB.TOTL.IN.ZS. Accessed Oct 2013
2. Gobierno De Chile. CONAMA. Plan De Descontaminacion De Santiago Chile. Marcelo Fernandez G., n. d. http://www.eclac.org/dmaah/noticias/paginas/3/32643/23-marcelo_fernandez.pdf
3. Gobierno De Chile. Ministerio Del Interior Y Seguridad Publica. Intendencia Metropolitana Declara Alerta Ambiental Preventiva Para Este Miércoles. Intendencia De Santiago. http://www.intendenciametropolitana.gov.cl/n4447_24-07-2013.html. Accessed 24 July 2013
4. Ministerio Del Medio Ambiente Del Gobierno De Chile, and Ministerio De Transporte Y Telecomunicaciones Del Gobierno De Chile (2012) E-Mobility Readiness Plan Chile (NAMA). Rep. Santiago: Gobierno De Chile (Print)
5. The World Bank (2012) GDP per Capita (current US$). World Bank National Accounts Data. http://data.worldbank.org/indicator/NY.GDP.PCAP.CD. Accessed Oct 2013

6. The World Bank (2012) Passenger Cars (per 1000 People) International Road Federation, World Road Statistics. http://data.worldbank.org/indicator/IS.VEH.PCAR.P3/. Accessed Oct 2013
7. Central Energía (2012) Power Plants in Chile. N.p. http://www.centralenergia.cl/en/power-plants-chile/. Accessed Oct 2013
8. Gobierno De Chile (2013) Comision Nacional De Energia. CNE Informa Nuevas Tarifas Electricas. N.p. http://www.cne.cl/noticias/energia/electricidad/752-cne-informa-nuevas-tarifas-electricas. Accessed Oct 2013
9. The World Bank (2012) Pump price for gasoline (US$ per Liter). Agency for International Cooperation. http://data.worldbank.org/indicator/EP.PMP.SGAS.CD/. Accessed Oct 2013
10. I-MiEV - El Futuro Es Manejable. Mitsubishi Chile. Mitsubishi Chile, n.d. http://www.mitsubishi-motors.cl/i-miev/. Accessed Oct 2013
11. Fuel Efficient Vehicle Tax Incentives Information Center (2013) US Department of Energy. http://www.fueleconomy.gov/feg/taxcenter.shtml. Accessed Oct 2013
12. Emprendedor Chileno Crea Automóvil Eléctrico Que Estará En Las Calles En Octubre (2013) La Segunda. http://www.lasegunda.com/Noticias/CienciaTecnologia/2013/06/853256/Emprendedor-chileno-crea-automovil-electrico-que-estara-en-las-calles-en-octubre. Accessed Oct 2013
13. Desafio Cero Homepage. Desafio Cero (n.d.) http://www.desafiocero.com/. Accessed Oct 2013
14. Ruta Verde Homepage. Ruta Verde, Chile (n.d.) http://www.rutaverdechile.cl/. Accessed Oct 2013
15. Chilectra: Electrolinera Y Puntos De Recarga. Chilectra (n.d.) http://www.chilectra.cl/wps/wcm/connect/ngchl/ChilectraCl/Hogar/TrabajandoComu/Proteccion del Medio Ambiente/Electrolinera. Accessed Oct 2013
16. Chilectra. Chilectra, Marubeni Y Petrobras Inauguran Primera "Electrolinera" De Carga Rápida De América Latina Para Autos Eléctricos. 20 Apr 2011. http://www.chilectra.cl/wps/wcm/connect/NGCHL/chilectracl/la+compania/comunicados+de+prensa/20110420+chilectra+marubeni+y+petrobras+inauguran+primera+electrolinera. Accessed Oct 2013
17. OECD Better Life Index: Chile (2012) OECD better life index. Organization for Economic Cooperation and Development (OECD). http://www.oecdbetterlifeindex.org/countries/chile/. Accessed Oct 2013
18. Ciclovías y Bicicletas Públicas. Sitio Oficial De La Municipalidad De Providencia. Municipalidad De Providencia, n.d. http://www.providencia.cl/servicios/ciclovias-y-bicicletas-publicas. Accessed Oct 2013
19. Ministerio Del Medio Ambiente (2013) Ministra Del Medio Ambiente Asistió Al Lanzamiento Del Primer Sistema Automatizado De Bicicletas Públicas En Chile. Gobierno De Chile, 13 Mar 2013. http://www.mma.gob.cl/1304/w3-article-53688.html. Accessed Oct 2013
20. Levinson DM, Ajay Kumar (1993) Multi-modal trip distribution: structure and application. Transp Res Rec 1466:124–31. (Print)
21. Instituto Nacional De Estadísticas. Censo de Chile (2002) Gobierno de Chile. http://www.ine.cl/. Accessed Oct 2013
22. Jaskula BW (2013) USGS mineral commodity summaries: lithium. United States Geological Survey, Jan. 2013. http://minerals.usgs.gov/minerals/pubs/commodity/lithium/mcs-2013-lithi.pdf. Accessed Oct 2013
23. Lithium Market (2012) Orocobre Limited. http://www.orocobre.com.au/Lithium_Market.htm. Accessed Oct 2013
24. Goonan TG (2012) Lithium Use in Batteries. USGS Circular no. 1371. United States Geological Survey. http://pubs.er.usgs.gov/publication/cir1371. Accessed Oct 2013

Lightning Source UK Ltd.
Milton Keynes UK
UKOW06n2318111115

262524UK00001B/17/P